ESSENTIAL FIBER CHEMISTRY

FIBER SCIENCE SERIES

Series Editor

L. REBENFELD

Textile Research Institute
Princeton, New Jersey

ESSENTIAL FIBER CHEMISTRY
by Mary E. Carter

CLOTHING: COMFORT AND FUNCTION
by Lyman Fourt and Norman R. S. Hollies

INSTRUMENTAL ANALYSES OF COTTON CELLULOSE
  AND MODIFIED COTTON CELLULOSE
by Robert T. O'Connor

Other Volumes in Preparation

ESSENTIAL FIBER CHEMISTRY

Mary E. Carter

FMC Corporation
American Viscose Division
Research and Development
Marcus Hook, Pennsylvania

MARCEL DEKKER, INC. New York   1971

A/677

MARCEL DEKKER, INC.
95 Madison Avenue, New York, New York 10016

LIBRARY OF CONGRESS CATALOG CARD NUMBER: 76-145882

ISBN NO.: 0-8247-1088-6

PRINTED IN THE UNITED STATES OF AMERICA

PREFACE

Each of the ten chapters in this book is concerned with a different natural or synthetic fiber encompassing the more commercially important fibers in the world today. For each fiber, the chemical and physical structure is briefly discussed including the fiber formation process as well as the chemical and physical properties. This is necessary background for the chemical modification of the fibers if one wishes to change or improve their properties. Properties such as dyeability, flammability, UV stability, soiling, static build-up can be altered by chemical means or by the addition of appropriate chemicals. A fiber's stress-strain properties can be altered, and the work recovery or crease recovery and abrasion resistance are some specific areas which can be modified.

Possibly the fiber's properties and modifications do not seem to be discussed in depth and with the detail that one might desire. This was unavoidable since a book could be written about each of these fibers. However, the present day modifications either as an after-treatment of the fiber or fabric or by addition of pertinent additives or comonomers in polymer formation are listed with appropriate references. The references for the most part were selected from the past ten years of textile chemistry but should be an adequate introduction to each fiber's history if one needs to pursue a specific problem. Some patent literature is included in order to give the reader an appreciation of product or process protection required in the development of today's fibers.

The properties of textile fibers are rapidly being changed and improved to become tomorrow's fibers. Tomorrow we might not immediately recognize an old fiber because of today's chemistry. These chapters are presented as a summation of the past and as a stimulant and guide for tomorrow.

The author is indebted to FMC Corporation, American Viscose Division for their support and the use of their technical information services. The assistance of Jacqueline Carter and Mildred Upton for their graphic art work, Dorothy Moore and Irma Kunkel for their assistance in the preparation of the manuscript is also appreciated. The advice and guidance of Ludwig Rebenfeld, Series Editor of Fiber Science Series, was invaluable in the finalizing of the manuscript.

MARY E. CARTER

January, 1971
Marcus Hook, Pennsylvania

CONTENTS

Contents

ESSENTIAL FIBER CHEMISTRY

Chapter 1 **COTTON**

I. Introduction

Of the several cellulosic fibers formed by nature, cotton is the one most extensively used in the world today. Due to their physical properties, other cellulosic fibers such as flax (linen), jute, hemp, and sisal have found certain specific end uses in today's market, but their production and consumption have not increased as has that of cotton. The cotton fiber is a single cell that grows as a seed hair on a plant belonging to the genus *Gossypium*. There are several species of commercial importance, principally, *G. hirsutum*, *G. barbadense*, and *G. arboreum*. A number of varieties and strains have been developed within each species and each has its identifying characteristics, both geometric and mechanical. The growth of the fiber occurs in several distinct stages and each stage adds to the formation of a complex, heterogeneous structure. The fiber's growth begins as a lengthening of an epidermal cell requiring about 13–20 days and produces a thin membrane called the primary wall. After the full length has been reached, further internal development occurs which results in fiber thickening as cellulose is deposited on the inner surface of the primary wall. This internal cellulose is called the secondary wall and never quite fills the cell fiber, leaving a center canal referred to as the lumen.

II. Structure

A. CHEMICAL

The cotton fiber is essentially cellulosic in nature and may be chemically described as poly(1,4-B-D-anhydroglucopyranose), with the following repeat unit:

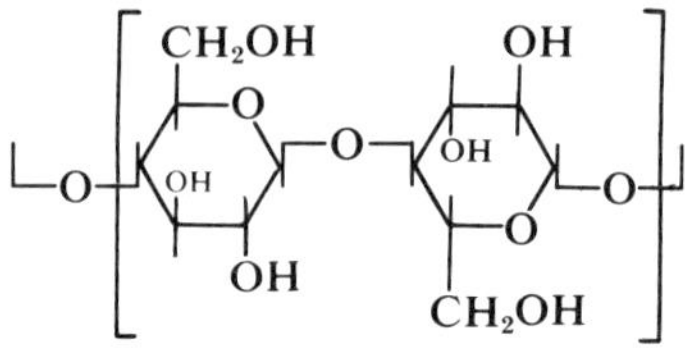

The degree of polymerization ($n/2$) of cellulose as found in cotton fibers is about 3000, but it may be higher in the native state since some unavoidable degradation probably occurs during isolation and purification. The roles that the primary and secondary hydroxyl groups and the β-glycosidic ether linkages play in defining the fiber's chemical properties and its finishing treatments are of prime importance. Formation of strong inter- and intramolecular hydrogen bonds and the natural stiffness of cellulose chains account for the fact that cotton is in a state of high crystallinity. Thus, for a better understanding of cotton fiber properties, we must consider in more detail how the cellulose is arranged at the supramolecular level.

B. PHYSICAL

1. *Supramolecular Structure*

In the development of the secondary wall, the cellulose is deposited in the form of fibrils. These fibrils in turn consist of microfibrils which contain polymer chains that are well packed and almost ideally oriented, and thus are crystalline. In the cotton fiber, these microfibrils are about $0.015\,\mu$ wide and approximately one third as thick. (Concentric lamellae are almost $0.1\,\mu$ thick.) The cellulose fibrils form a helix in the fiber and the angle of the helix, with respect to the fiber axis, is constant both cross sectionally and along the length of the fiber. In the most common cottons the helical angle of the fibrils is about 35°, as determined from an angular measurement of the cellulose 002 diffraction arc(*1*). The fibrillar nature of the secondary wall has been clearly shown by Rollins and Tripp (*2*) in electron photomicrographs.

The fibrils in the fiber form in laminar sheets or growth layers which are associated with the day–night cycles during growth and the corresponding variations in temperature, internal moisture content, relative humidity, and light intensity. Work continues in this area, for there are other variables and growth layer characteristics which are not fully understood (*3*).

A unique feature of cotton fiber morphology is that the sense of the helical deposition of the fibrils in the secondary wall reverses from a left to right direction, and vice versa, many times along the length of the fiber. These structural reversals may be considered as points of weakness, since under uniaxial tension, cotton fibers break preferentially at points of reversal rather than between them. On the other hand it is because of these reversals that cotton fibers have extensibilities on the order of 8% despite their high crystallinity and high levels of fibrillar orientation with respect to the fiber axis. It should be noted that cotton is the only naturally occurring cellulosic fiber that has helical reversals (*4*).

2. *Fine Structure*

The unit cell dimensions and index of refraction of crystalline cellulose have been found to change depending on the chemical and thermal history of the polymer. There are four unit cell structures recognized today, and these are designated as Cellulose I, II, III, and IV. Cellulose I is the crystalline cellulose as found in an untreated cotton fiber; Cellulose II is produced when Cellulose I is treated with strong sodium hydroxide (18–20%). Cellulose III may be formed by treating cotton (Cellulose I) with anhydrous ethylamine (*5*). Cellulose IV, a unit cell whose existence has caused considerable discussion, can be made by treating ethylene–diamine complexes of Cellulose I, II, or III with dimethylformamide at 160°C (*6*), or it has been made from Cellulose III in glycerol at 250°C (*7*).

The latest picture of the unit cell is the hydrogen-bonded network described by Liang and Marchessault (*8*) in which the cellulose chains are depicted as being parallel and antiparallel. The intermolecular hydrogen bonds are described as (1) between C_6 hydroxyls of antiparallel chains and glycosidic ether oxygens of parallel chains, (2) between C_6 hydroxyls of parallel chains and glycosidic ether oxygens of antiparallel chains, and (3) between C_6 and C_2 hydroxyls on adjacent chains. There is also an intramolecular hydrogen bond between the C_3 hydroxyls and the ring oxygens of adjacent anhydroglucose units of the same chain (see Fig. 1).

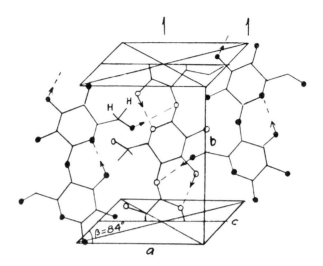

FIG. 1.   The Liang–Marchessault unit cell for cellulose.

III.   Properties

A. PHYSICAL

Cotton fibers vary in length, fiber diameter (fineness), strength, elongation, convolution frequency, and maturity. These variations are due to genetic factors and to environmental growth conditions such as rainfall, humidity, sunlight, nutrients, and conditions and methods of harvesting. Within any one bale in any one year, the individual fibers display a broad distribution of physical properties. Fiber properties fall into the general ranges given in Table 1.

TABLE I

Properties		Range
Fiber length		$\frac{7}{8}-1\frac{7}{16}$
Fineness (weight/in., μg)		2.7–5.5
Tenacity, g/den.	dry	3.0–4.9
	wet	3.3–6.4
Elongation, %		4–13
Specific gravity		1.54–1.56
Moisture regain (70°F, 65% r.h.)		7–8

Cotton fibers, from the general discussion of the supramolecular structure, have a high degree of fibrillar orientation which accounts for their good strength and low elongation. Rebenfeld (9) has correlated the fibrillar orientation with these and other physical and chemical properties of cotton fibers.

B. CHEMICAL

The cellulose molecule is subject to hydrolysis by acid. This hydrolytic attack occurs at the β-glucosidic linkage and results in chain scission which produces an additional reducing end. The rate and extent of hydrolysis depends on the type of acid (organic or inorganic), concentration of the acid, temperature, and accessibility of the cellulose to the reactant. In fact, several earlier methods for determining the degree of crystallinity were based on rate and extent of acid hydrolysis under carefully controlled conditions. These procedures are subject to criticism since increases in crystallinity can occur after initial chain scission in the accessible areas.

The chemical reactions between cellulose and alkali are more varied and reaction pathways are dependent upon the presence or absence of oxygen. Of lesser importance commercially is the degradation that occurs in the absence of oxygen. This reaction is not random and begins at the reducing end of the polymeric chain. In a "peeling-type" reaction, saccharinic acids are produced stepwise from each glucose unit with D-glucoisosaccharinic acid predominating. In the presence of oxygen, the alkaline degradation reaction may more aptly be described as the oxidation of alkali cellulose. In this instance, the observed result is random chain scission as shown by a lowering of molecular weight. This reaction is of importance in the aging process in the manufacture of rayon, and a more detailed discussion can be found in the chapter on that fiber.

Other chemicals are also capable of oxidizing cellulose and, in general, they are unspecific in their action. The exceptions to this statement are periodic acid and its salts and nitrogen oxides, particularly dinitrogen tetroxide. Periodic acid and its salts oxidize the adjacent C_2 and C_3 secondary hydroxyls to aldehyde groups, while dinitrogen tetroxide oxidizes predominately the C_6 primary alcohol group to a carboxyl group. Reaction with N_2O_4 also produces some reducing groups as well as nitrogenous ester groups. The three hydroxyl groups on each glucose moiety in the polymer chain are subject to oxidation according to the following general scheme:

The hydrolysis of the dialdehyde yields glyoxal and o-erythrose.

Other properties of cotton fibers are that they burn readily, are hydrophilic and swell in water, and are readily attacked by certain fungi and bacteria. The uv portion of sunlight promotes oxidation by atmospheric oxygen resulting in "oxycelluloses" as described above and a lowering of fiber strength. Gamma irradiation of cellulose results in depolymerization and a reduction in crystallinity.

IV. Finishing

A. SCOURING AND BLEACHING

Cotton fiber, while consisting principally of cellulose, also contains small quantities of protein, pectic substances, waxes, organic acids (such as malic and citric), simple sugars, and traces of metal salts. The noncellulosic substances include approximately 0.5% oil and wax, 1%

nitrogen-containing compounds, and 1–1.8% mineral salts (*10*). These impurities are removed commercially by processes broadly termed "scouring and bleaching," and usually these treatments are carried out on the fabric. Since the warp yarns of the fabric have been coated (sized) with a starch or other film-forming material containing a lubricant, and frequently a humectant, in order to reduce friction in the weaving process, the fabric must be desized either with suitable enzymes or by steeping in dilute acid. After washing, the fabric is then given a hot, dilute alkali treatment to remove a major percentage of the waxes. These treatments are mild and the actual molecular weight reduction is small. The bleaching process completes the removal and oxidation of the remaining colored impurities in the fabric. The bleaching process may utilize either a chlorine or a peroxide-type oxidizing agent.

The chlorine-type bleaches include, principally, sodium hypochlorite and sodium chlorite, although some commercial application has included sodium bromite, chlorine dioxide, and chlorine. Bath stability and evolution of gaseous chlorine compounds necessitate careful control of pH in the bath. This is normally accomplished with sodium carbonate, although recent work shows that sodium dihydrogen phosphate is very effective (*11*).

In the past several years, bleaching with peroxy compounds has received a great deal of attention and commercial acceptance. Hydrogen peroxide and peracetic acid are frequently mentioned, but again careful control, particularly of pH, is required. Bleaching with hydrogen peroxide is carried out at alkaline pH, thus allowing scouring and bleaching to take place concurrently. In acid or neutral solutions the liberation of perhydroxyl ions is too slow, and in excess alkali the hydrogen peroxide is unstable and evolves oxygen. Bath stability is easily accomplished with magnesium or sodium silicate; however, the possible deposition of silicates on fabric and machinery, particularly in package dyeing, is undesirable. Usually, this can be controlled by adding more alkali in the form of equal portions of sodium hydroxide and sodium carbonate.

B. MERCERIZATION

In 1850 John Mercer found that treatment of cotton with a caustic soda solution caused a diametral swelling of the fiber and a longitudinal shrinkage. A short time later Horace Lowe (1889–1890) showed that a high luster could also be obtained if the yarn or fabric was held under tension during the alkali treatment. Today, the term mercerization means the treatment of cotton with a strong sodium hydroxide solution, slack or under varying tension. The caustic concentration may range between 12 and 25% and the temperature is usually decreased as the caustic concentration is lowered.

General procedures for mercerization vary depending on the objective to be achieved. A typical procedure includes complete wetting of the cotton in the form of yarn or fabric with a caustic soda solution (usually 18–25% sodium hydroxide) for 1–4, min at 35–40°C. The caustic soda is removed by washing with water at 50–60°C followed by a dilute acid wash (1–3% AcOH, H_2SO_4). The cotton is then freed of acid by a water wash and the last water wash may contain an alkaline softener to neutralize the last traces of acid. The cotton fabric is then dried at 110–115°C. Mercerization is usually carried out under some tension to control shrinkage both in length and width of the fabrics. When open-weave fabrics are slack mercerized allowing excessive shrinkage, a "stretch" fabric results with semielastic properties.

Mercerization changes the unit cell of the cotton crystallites from Cellulose I to Cellulose II, and is an irreversible process. The caustic treatment also reduces the degree of crystallinity from about 70 to 50% and decreases the size of the crystallites. In addition, the physical property changes observed in the fibers are (1) increased moisture absorption, (2) increased dye absorption, (3) increased circular cross section, and (4) increased cross sectional area. If the yarn or fabric is mercerized under tension or restretched before washing, increased strength and luster are obtained. If the yarn or fabric is mercerized slack, increased elongation at break and increased apparent elastic recovery with decreased length are the results. There are small decreases in fiber density reflecting the decreased crystallinity. The extent of these property changes also depends on the characteristics of the original fiber, construction of the yarn (twist and denier), and construction of the fabric.

Mercerization procedures may be carried out on fabric in the gray state or after scouring and bleaching. Regardless of the point of treatment it is important to obtain a thorough, uniform penetration of the fabric and yarn, and wetting agents are frequently used, particularly on gray goods.

C. CHEMICAL MODIFICATION VIA CROSSLINKING REACTIONS

When the first patent applications were filed by Tootal, Broadhurst, and Lee over 40 years ago for shrink resist and noncrush cottons by condensation *in situ* of resins, such as urea–formaldehyde and phenol–formaldehyde, little was known about the molecular processes involved. Now it is widely accepted that these fabric properties are achieved by crosslinking the cellulose chains through the hydroxyl groups on the glucose units. Considerable effort has been expended in finding appropriate crosslinking agents, and recently, some consideration has been given to the importance of locating the crosslink in the cellulose molecule. There are many agents available and they tend to be either monomeric compounds or low molecular weight polymers with two or more reactive sites. They react with the hydroxyl groups, preferably on different cellulose molecules, at elevated temperatures with the aid of catalysts, usually by a condensation mechanism. The classical procedure for their application is the pad-dry-cure process, where the drying is accomplished at around 100°C, and the curing, during which most of the crosslinking occurs, at 135–165°C.

Recently there has been a growing commercial interest in the use of radiation techniques in the chemical modification of textiles, particularly with the acceptance of durable press qualities. The economies of this approach must be worked out for each radiation process since the energy source may be x rays, electrons, neutrons, and alpha and other particles of sufficient energy to cause ionization. About 20 eV with radiation is required to break a chemical bond, although its strength is more like 5 eV. If the source of energy to be used is either β rays (high-energy electromagnetic waves) or γ rays (high-energy electrons), then 300 keV to 1 MeV need be considered, and although the two forms of radiation are different, their interaction with matter produces similar effects. The application of radiation energy for the "cure" in permanent-press fabrics treated with N-methylolacrylomide has been reported by Walsh, Jin, and Armstrong(*12*).

1. Crosslinking Agents

The oldest and most important group of crosslinking agents is formed by reacting formaldehyde with diamino and diamido compounds to form diamino- and diamidomethylol adducts. These react with hydroxyl groups in the presence of acidic catalysts at elevated temperatures. Free acids, such as lactic, tartaric, and hydrochloric, or latent Lewis acids, such as zinc nitrate and magnesium chloride, may be used. Other catalysts which have been studied and used, include zinc fluoroborate, ammonium sulfate, ammonium phosphate, and amine hydrochlorides. The crosslinking agents include:

(1) Methylolureas

example:

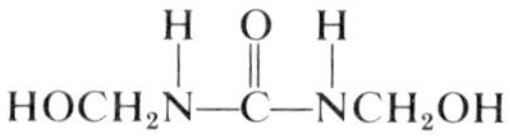

$$\text{HOCH}_2\overset{\overset{\text{H}}{|}}{\text{N}}\text{—}\overset{\overset{\text{O}}{\|}}{\text{C}}\text{—}\overset{\overset{\text{H}}{|}}{\text{N}}\text{CH}_2\text{OH}$$

(2) Cyclic alkyleneureas *(13)*

example:

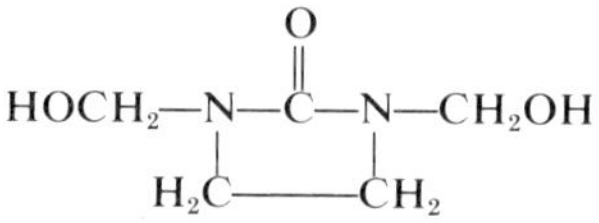

$$\text{HOCH}_2\text{—N—}\overset{\overset{\text{O}}{\|}}{\text{C}}\text{—N—CH}_2\text{OH}$$
$$\text{H}_2\text{C}\text{———CH}_2$$

1,3-bis(hydroxymethyl)-2-imidazolidinone (DMEU)

(3) Triazone

example:

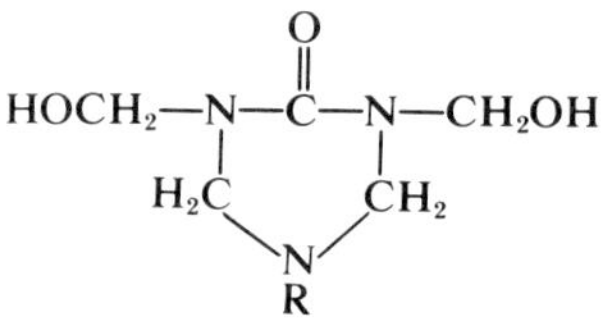

$$\text{HOCH}_2\text{—N—}\overset{\overset{\text{O}}{\|}}{\text{C}}\text{—N—CH}_2\text{OH}$$
$$\text{H}_2\text{C}\qquad\text{CH}_2$$
$$\underset{\text{R}}{\text{N}}$$

(4) Uron

example:

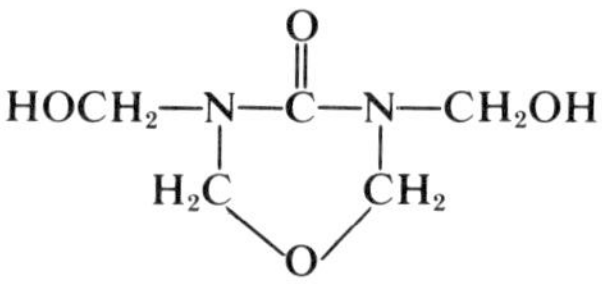

$$\text{HOCH}_2\text{—N—}\overset{\overset{\text{O}}{\|}}{\text{C}}\text{—N—CH}_2\text{OH}$$
$$\text{H}_2\text{C}\qquad\text{CH}_2$$
$$\text{O}$$

(5) Triazines

$$\text{H}_2\text{O—CH}_2\text{—N—H}$$

example:

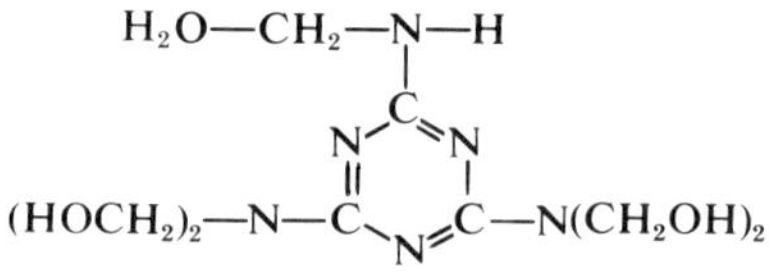

$$\overset{\text{C}}{\underset{\text{N}}{\text{N}}}\overset{}{\underset{}{\text{N}}}$$
$$(\text{HOCH}_2)_2\text{—N—C}\underset{\text{N}}{\underset{}{}}\text{C—N(CH}_2\text{OH)}_2$$

(6) Carbamates

example:

$$\text{HOCH}_2\text{—}\underset{\underset{\text{H}}{|}}{\text{N}}\text{—}\overset{\overset{\text{O}}{\|}}{\text{C}}\text{—O—CH}_2\text{CH}_2\text{O}\overset{\overset{\text{O}}{\|}}{\text{C}}\underset{\underset{\text{H}}{|}}{\text{N}}\text{CH}_2\text{OH}$$

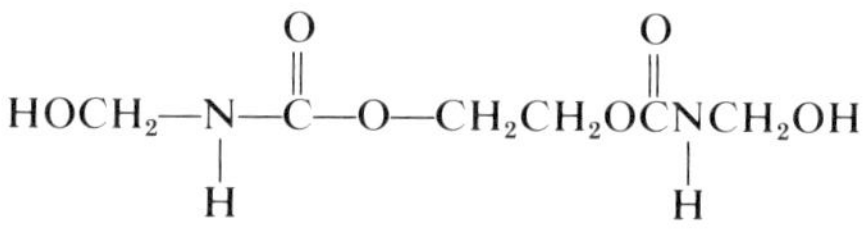

(7) Methylolamides

example:

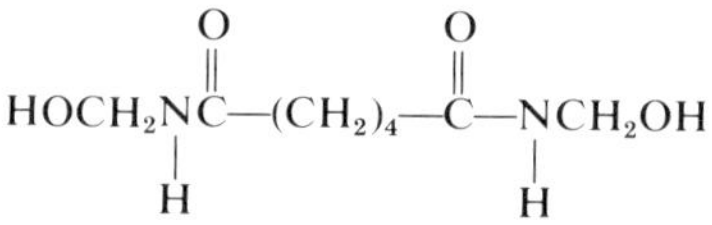

$$\text{HOCH}_2\text{N}\overset{\overset{\text{O}}{\|}}{\text{C}}\text{—(CH}_2)_4\text{—}\overset{\overset{\text{O}}{\|}}{\text{C}}\text{—NCH}_2\text{OH}$$
$$\qquad\underset{\text{H}}{|}\qquad\qquad\qquad\qquad\underset{\text{H}}{|}$$

Other crosslinking agents include the acetals, sulfone derivatives, and aziridine derivatives. Some of the crosslinking agents require rather strong Lewis acids and, in general, catalysis of the methylol group is related to the basicity of the moiety to which it is attached.

These compounds react with cotton fiber in the following manner:

$$2\,\text{Cell OH} + \text{R}(\text{CH}_2\text{OH})_2 \longrightarrow \text{Cell-O}-\text{CH}_2\text{RCH}_2\text{O-Cell} + 2\,\text{H}_2\text{O}$$

The earliest reactions with urea–formaldehyde and melamine–formaldehyde resins were probably twofold with the methylol groups ($-\text{CH}_2\text{OH}$) reacting both with the cellulose, as shown above, as well as reacting with itself to yield a three-dimensional resin. It is generally accepted that the desirable effects are achieved by the introduction of covalent crosslinks and not by the internally deposited polymer resulting from the self-condensation reaction of the dimethylol crosslinking agent. All aspects of this question have not been resolved, but it has been shown that only a small fraction of the reagent which is added to the fiber is involved in covalent crosslinking.

The first resin finishes and crosslinking reactions changed the crease recovery angle of a cotton fabric from approximately 160° to 200–220°. The wash-and-wear era of the 1950's required angles of 230–260° and the fabrics were noted for their smooth drying qualities. Now, in the era of durable press, a crease recovery angle of 280–320° is required. The concept of durable press is based on the application and fixation of the reactant(s) to the fabric with curing deferred until the garment is fabricated. This transfers the crosslinking reactions from the finishing plant to the garment manufacturer. Here, the appropriate creases and pleats are put in, and the garment is cured for a predetermined time and temperature.

Many scientists have contributed information to the crosslinking processes for cotton, and to the durable press garments of today. These processes were introduced in patents to Warnock and Hubener (*14*), Buck and Getchell(*15*), and Hurwitz(*16*). Today there are a variety of durable-press processes which Hepp(*17*) has divided into three groups: (1) postcuring, (2) precuring, and (3) two-stage curing. In all these processes, the chemical reactions are similar but the distribution and number of crosslinks vary. In some instances, there may be a balance achieved between crosslinks and resin formation. The variables, which are used to control the extent of crosslinking and its distribution, are moisture content and/or swelling, reaction temperature, reaction time, nature of the catalyst, and reactivity of the reagent.

The crosslinking reaction introduces both desirable and undesirable fabric properties. The cotton fiber may be viewed as a highly crosslinked fiber in its native state due to the network of hydrogen bonds and high crystallinity. The addition of more crosslinks through covalent bonding tends to embrittle the fiber and, unlike the hydrogen bonds, these covalent crosslinks are stable under wet conditions and at elevated temperatures. The introduction of crosslinks yields a fiber with higher resilience, defined as recovery from deformation, decreased extensibility, and decreased strength. Such changes in the fiber are consistent with a decreased chain mobility which is associated with crosslinking. In fabrics these changes appear as improved wrinkle recovery and crease resistance but decreased fabric strength and abrasion resistance. Softeners such as an emulsified polyethylene provide fiber lubrication and suppress some of the abrasive wear. A stable polyurethane emulsion has been used in conjunction with the softener, though it can only be used on colored fabrics (*18*).

2. *Reactivity and Accessibility*

a. *Formaldehyde Studies.* In the reaction between cellulose and a variety of difunctional reagents, we can expect the reactivities of the hydroxyl groups in the C-2, C-3, and C-6 positions of the glucopyranosyl repeat unit to be influenced by the electron densities at the individual oxygen atoms, the steric interferences around the hydroxyl groups, the participation of the hydroxyl groups in hydrogen bonding, and the nature of the reactant. In addition to reactivities as influenced by these chemical factors, the availability of the hydroxyls in less ordered or amorphous regions and on the surface of the ordered or crystalline regions of the fiber is an important factor in the extent of chemical modification which is achieved and, therefore, in the observed changes in physical fiber and fabric properties. The accessibility of the hydroxyl groups are dependent on the nature of the reagents and the conditions under which they are applied and reacted.

The regions of a cotton fiber which undergo reaction during a chemical modification procedure have been shown by Rollins and coworkers (*19–21*) via electron microscopy. By examining fiber fragments and thin cross sections after treatment with cellulose solvents which dissolve the unreacted cellulose, one can locate the crosslinked regions. Observation of crosslinked cottons in this way showed that the periphery of the fiber may be crosslinked before the center, and that this results in poor physical properties. Fibers, uniformly reacted in the wet or swollen state, can swell extensively in a solvent but will not dissolve. Such samples will have better wet recovery angles than dry. When fibers have good dry crease recovery, the observed cross sections would neither swell nor dissolve.

Using formaldehyde as a model crosslinking reagent, Rowland and Post(22) set out to characterize the network structure in modified cotton by defining two basic parameters: (1) frequency or concentration of crosslinks and (2) distribution of crosslinks. Crosslinking with formaldehyde under a variety of conditions and employing a solution method of removing the uncrosslinked or soluble cellulose, Rowland obtained sol–gel fractions which showed that the most efficient utilization of HCHO for insolubilization occurred in an aqueous process, and the least efficient utilization occurred in a nonaqueous process. In studying the distribution of the crosslinks by electron micrographic analysis and by kinetic methods, rapid and slow phases of the reaction were evident which varied with the specific processes. A broad range in heterogeneity of distribution of crosslinks was found with the most nonuniform distribution appearing in a high concentration of crosslinks in the peripheral regions of the fiber. The conclusion was reached that relatively low heterogeneity of distribution of crosslinks results from a typical pad-dry-cure process where time for diffusion of the reagent is allowed before reaction. When reaction proceeds with immersion and penetration of the reactant into the fiber, the available sites involved are in those regions which can be reached through swelling and diffusion processes. Thus, the greater the difference between the reaction rate and the diffusion process, the more accentuated is the nonuniformity of the crosslink distribution(23, 24).

When dried cotton is exposed to formaldehyde vapor and there are no changes in the relative amounts of crystalline and accessible regions during the reaction, only 65–95% of the potentially accessible hydroxyl groups are reached in the vapor phase reaction(25). There are many methods which have been used to determine accessibility, including moisture regain, infrared deuteration, formylation, iodine sorption, acid hydrolysis, and periodate oxidation. When Jeffries, Roberts, and Robinson(26), using these techniques, examined a series of cotton and partially methylated cotton fabrics, they concluded in the light of the morphology of the cotton fiber, that the accessibility of the hydroxyls on the surfaces of the elementary fibrils amount to 40–70% of all the hydroxyl groups.

The reactivities of the three hydroxyl groups in the D-glucopyranosyl repeat unit of cellulose have been summarized recently by Rowland et al.(27). Based on the research of many scientists, it is evident that various reagents distribute themselves in significantly different ratios among the available hydroxyl groups in the glucopyranosyl units of cellulose. For rate-controlled reactions in basic media, such as in a

Williamson synthesis and closely related irreversible reactions, the C-2 and C-6 hydroxyl groups are predominantly involved, whereas in equilibrium-controlled reactions, the C-6 hydroxyl group preferably is involved. The only known exception to this is the base-catalyzed reaction of ethylene oxide which is irreversible and occurs predominantly at the C-6 primary hydroxyl.

The distribution of substituents among the hydroxyl groups at C-2, C-3, and C-6 has been shown to be dependent upon the reagent(28) and the conditions of the reaction(29, 30). Reactions which occur in media that are weaker swelling agents than mercerizing strength sodium hydroxide take place on the surface of microstructural units of cellulose in the cotton fiber(31, 32). In addition, there is a further selective accessibility for the individual hydroxyls on these microstructural units.

The conclusion of Willard, Turner, and Schwenker(33), upon analysis of a trimethylolmelamine-treated cotton, and Patel and coworkers(34) upon analysis of a formaldehyde-crosslinked cotton, is that the order of decreasing reactivity of the hydroxyl groups is C-6 > C-2 > C-3. These reactions were carried out by pad-dry-cure processes with acid catalysis. Therefore, in these systems there seems to be, as would be expected, a selective accessibility and reactivity for the individual hydroxyls on the microstructural units of the cotton fiber.

b. *Kinetic Studies.* Benerito and coworkers(35–39), measured the reaction rates of a series of urea derivatives which are capable of crosslinking cellulose and producing high wet and dry crease recovery angles. The reactions were catalyzed with zinc nitrate, zinc chloride, magnesium nitrate, and magnesium chloride at several temperatures.

For at least 50% of the completed reaction, all data showed pseudo first-order behavior with the exception of the reaction of DMeDHEU catalyzed with the magnesium salts; this appears to be a pseudo zero-order reaction. The reaction rate constants are listed in Table II. Examination of these values shows that the order of decreasing reactivity of the substituted ureas with cotton is as follows:

DMEU $\gtrsim$ DMPU > DMeDHEU > DHEU $\gtrsim$ DMDHEU.

$$
\begin{array}{c}
\text{O} \\
\| \\
\text{HOH}_2\text{CN}\text{---}\text{C}\text{---}\text{NCH}_2\text{OH} \\
|\quad\quad\quad\quad| \\
\text{H}_2\text{C}\text{------------}\text{CH}_2
\end{array}
$$

1,3-bis(hydroxymethyl)-2-imidazolidinone (dimethylolethyleneurea, DMEU)(35)

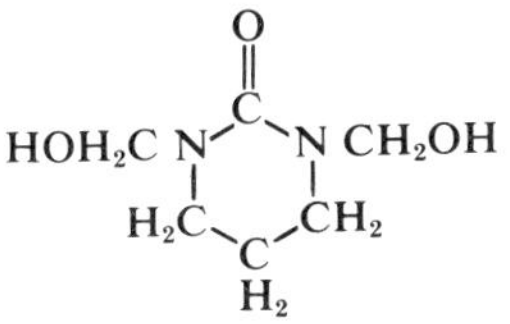

$$
\begin{array}{c}
\text{O} \\
\| \\
\text{C} \\
\end{array}
$$

HOH$_2$C N ⟶ N CH$_2$OH

1,3-bis(hydroxymethyl)-2-(1H)-tetrahydropyrimidinone
(dimethylolpropyleneurea, DMPU)(*39*)

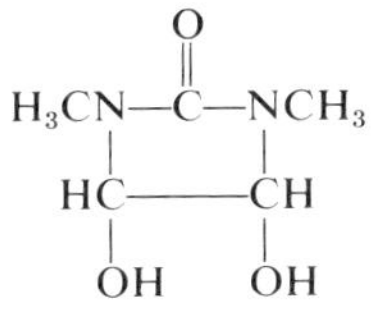

H$_3$CN—C—NCH$_3$
HC——CH
OH   OH

1,3-dimethyl-4,5-dihydroxy-2-imidazolidinone
(dimethyldihydroxyethyleneurea, DMeDHEU)(*38*)

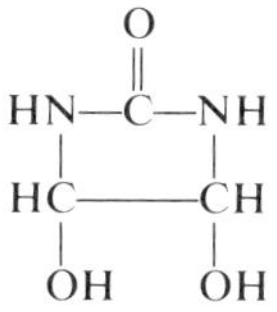

HN—C—NH
HC——CH
OH   OH

4,5-dihydroxy-2-imidazolidinone (dihydroxyethyleneurea, DHEU)(*36*)

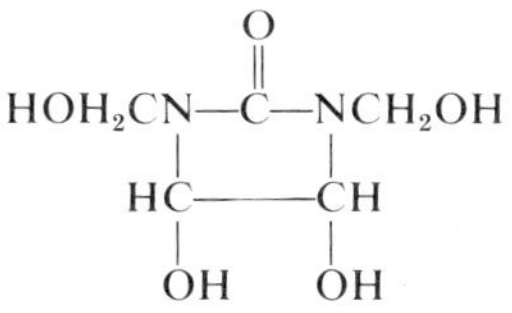

HOH$_2$CN—C—NCH$_2$OH
HC——CH
OH   OH

4,5-dihydroxy-1,3-bis(hydroxymethyl)-2-imidazolidinone
(dimethyloldihydroxyethyleneurea, DMDHEU)(*37*)

From infrared observations, the preferential cellulose reaction sites
are indicated as given in Table III.

The reaction rates and infrared data show that the methylolated
derivative of DHEU (DMeDHEU) has decreased reactivity of the
methylol group, due probably to the presence of the ring hydroxyls,
particularly at low temperatures. The methylation of the amido groups
of DHEU shows increased reactivity of the ring hydroxyls with zinc
salts. Since the reaction rates for DMEU and DMPU were similar, it
would appear that increasing the ring size from five to six atoms had no
significant effects. Yet, infrared absorbencies do indicate differences in
the type of reaction occurring and there is a difference in the N/HCHO
ratio for the two reactants.

TABLE II

Specific Reaction Rate Constants at 0.03 M Catalyst Concentration

$k \times 10^5$ min^{-1}

Catalysts	Temp.	DMEU based on		DHEU based on		DMDHEU based on		DMeDHEU based on	DMPU based on	
		N.%	HCHO.%	N.%	HCHO.%	N.%	HCHO.%	N.%	N.%	HCHO.%
None (reactant alone)	45	1.9	2.5	1.1		0.4	0.4	1.3	3.8	2.4
	55	4.3	3.4	2.5		0.5	1.2	0.7	8.4	9.1
	65	11.6	8.1							
MgCl$_2$	45	8.8	9.4	0.3		0.7	0.4		5.1	5.7
	55	32.1	20.6	3.3		1.4	1.5		12.2	13.1
	65	50.9	37.0	9.5		6.1	4.0		59.2	43.1
	75			40.9		67.4	42.1			
	85			83.4		209.9	206.4			
Mg(NO$_3$)$_2$	45	13.7	9.9	0.4		1.0	0.7		14.2	3.4
	55	31.1	29.9	5.2		1.6	2.8		20.5	14.5
	65	52.0	31.6	10.2		6.8	3.8		32.5	40.2
	75			38.7		36.5	27.0			
	85			97.4		171.1	126.4			
ZnCl$_2$	45	67.2	27.0	2.2		2.1	1.2	7.8	65.9	60.1
	55	63.3	89.2	23.7		5.5	2.2	38.2	107.4	106.0
	65	229	335	25.0		6.8	11.0	126.0	419.8	449.3
	75			597.9		216.7	126.7			
	85			884.3		638.6	372.3			
Zn(NO$_3$)$_2$	45	444	288	8.8		4.5	8.8	152.0	377.3	260.4
	55	519	397	39.1		22.9	31.7	1133.0	743.2	543.3
	65	640	397	166.0		39.7	123.1	4360.0	4201.1	1452.0
	75			1714.6		1675.9	898.2			
	85			3252.8		7804.9	6973.5			

TABLE III

Reactant	Preferential reaction site on cellulose
Dimethylolethyleneurea	primary hydroxyls
Dihydroxyethyleneurea	secondary hydroxyls
Dimethyloldihydroxyethyleneurea	
(a) methylol groups	primary hydroxyls
(b) ring hydroxyl	secondary hydroxyls
Dimethyldihydroxyethyleneurea	—

Some kinetic data for linear crosslinking agents (40), the dimethylol carbamates, have been obtained at a cure temperature of 160°C. With as much as 80% of the reaction completed, the pseudo first-order reactions shown in Table IV were observed.

TABLE IV

$\begin{matrix} O \\ \parallel \\ ROC\!-\!N(CH_2OH)_2 \\ \mid \\ R \end{matrix}$	Specific reaction rate constant $(k \times 10^2)$ sec^{-1}
CH_3^-	5.2
$C_2H_5^-$	3.7
$HOC_2H_4^-$	3.5
$ClC_2H_4^-$	2.5
$CH_3OC_2H_4^-$	2.3

While these kinetic studies clarify many of the empirical observations made in the past, there are still many complexities that require further inquiry. For example, the infrared data of Benerito and coworkers suggest a metal ion to nitrogen bonding for DMEU, but not for DMPU, while with DHEU the possible sites for coordination are the amide groups and the pendant ring hydroxyls. With DMeDHEU, the metal ion appears symmetrically bonded to the π-electron cloud of the amide group. From data obtained by a precipitation method, Haith and Namboori (41) have suggested that the complex formation is possible between the carbonyl oxygen and the methylol hydroxyl rather than the nitrogen. The compounds they studied were dimethylolurea, N-methylolacrylamide, and dimethylol ethyl triazone. This suggestion based on their results is valid upon consideration of Penland and coworkers' (42) infrared data of metal–urea complexes. The important point, however, is that Benerito's observations were made on cyclic urea compounds, capable of resonant hybrid structures, such as

$$
\begin{array}{ccc}
& & \mathrm{NH-CHOH} \\
& \diagup & | \\
\mathrm{-O-C} & & \\
& \diagdown & | \\
& & \mathrm{NH-CHOH} \\
& & +
\end{array}
\qquad
\begin{array}{ccc}
& & \overset{+}{\mathrm{NH}}\mathrm{-CHOH} \\
& \diagup & | \\
\mathrm{-O-C} & & \\
& \diagdown & | \\
& & \mathrm{NH-CHOH}
\end{array}
$$

and if coordination occurs between $>C{=}O{\rightarrow}$ metal then bands in the $6\text{-}\mu$ region should shift to lower frequencies and not to the higher frequencies, as was observed by Benerito. Thus, the substituent groups on the urea molecule can be very influential in determining the preferred coordination of the metal ion.

From the temperature dependence of the rate constants, the enthalpies, entropies, and free energies of activation have been calculated for the various urea derivatives, and these are shown in Table V. Probably, the variations in entropy, ΔS, are the most interesting, for this reveals some information about the transition state complex of the metal ions. The greater the entropy change, the greater the increase in disorder of the transition state complex over that of the reactants and the more "loosely bound" is the complex. A large negative value, such as those obtained for DMEU, is indicative of an S_{n^2} mechanism or a highly ordered transition state complex. From the free energy of activation, the reactivities of the reactants are obtained and, as previously stated, DMEU reacts fastest and is associated with the lowest energy for activation.

Throughout the studies by Benerito, it was noted from infrared measurements that the absorbed water (band, $6.1\ \mu$) of native cellulose always disappeared in crosslinked fabrics that exhibited high dry crease recoveries. With the present-day durable-press processes, it was suggested that this band might be used as a guide for establishing the extent of reaction in operations involving partial or delayed cures.

V. Other Finishing Treatments

A. DYEING

The dyeing of cotton in fabric or fiber form is accomplished by three principle processes. Cotton may be chemically reacted with dyestuffs in solution; these compounds are known as fiber-reactive dyes. A second method is the use of substantive dyes which diffuse

TABLE V

ENTHALPIES, ENTROPIES, AND FREE ENERGIES OF ACTIVATION AT 45°C OF CELLULOSE CROSS-LINKING AGENT REACTIONS IN THE PRESENCE OF 0.03 M INORGANIC SALT CATALYSTS

		ΔH, kcal/mole		ΔS, cal/mole		ΔF, kcal/mole	
		N.%	HCHO.%	N.%	HCHO.%	N.%	HCHO.%
DMEU	$MgCl_2$	19.3	15.2	−26.6	−39.4	27.7	27.7
	$Mg(NO_3)_2$	14.8	14.7	−40.0	−40.6	27.5	27.0
	$ZnCl_2$	13.6	27.8	−40.5	+ 2.7	26.1	27.0
	$Zn(NO_3)_2$	4.0	3.5	−66.8	−69.2	25.3	25.6
DHEU	$MgCl_2$	34.5		18.4		28.7	
	$Mg(NO_3)_2$	34.9		22.3		27.8	
	$ZnCl_2$	31.4		2.8		30.5	
	$Zn(NO_3)_2$	29.6		0.1		29.6	
DMDHEU	$MgCl_2$	34.5	35.7	16.1	18.7	29.4	29.7
	$Mg(NO_3)_2$	30.2	28.5	3.2	− 2.9	29.2	29.4
	$ZnCl_2$	34.0	35.0	16.7	18.7	28.7	29.1
	$Zn(NO_3)_2$	43.3	37.6	47.5	30.8	28.2	27.8
DMeDHEU	$MgCl_2$	20.7		−26.3		29.1	
	$Mg(NO_3)_2$	16.7		−39.2		29.2	
	$ZnCl_2$	29.8		6.0		27.9	
	$Zn(NO_3)_2$	35.9		31.2		26.0	
DMPU	$MgCl_2$	26.1	21.6	− 6.3	−20.4	28.1	28.1
	$Mg(NO_3)_2$	18.8	26.4	−58.6	− 6.1	27.4	28.3
	$ZnCl_2$	19.7	21.4	−21.5	16.3	26.5	26.6
	$Zn(NO_3)_2$	25.6	18.3	0.65	−23.0	25.4	25.6

directly into the fiber from a dye solution. The third method is referred to as mordant dyeing in which the dyestuff in solution reacts with metals previously applied to the fiber to form an insoluble colored compound on the cotton. Dyestuffs, in general, depend on unsaturated groups such as $-N=N-$, $=C=O$, $=C=S$, $=C=N-$, $-N=O$, and $=C=C=$ for their colors. These chromophores are found in the chemical structure of dyestuffs although the dye manufacturer does not always reveal the complete chemical formula.

Fiber reactive dyes(43) for cotton may be represented by a nitrogen-containing heterocyclic compound having a halogen and one or two groups containing the chromophore, such as this derivative of cyanuric chloride. The dyestuff reacts with the cellulose hydroxyl at the halogen

Y is the group containing chromophore

site. Among the substantive dyes are the simple direct dyes which diffuse into the fiber and whose dyeing rate is increased by the addition of electrolytes. Developed direct dyes are produced by impregnation with the dye, diazotization of the amino groups, and coupling with a naphthol to produce a new dye on the fiber. Vat dyes are another important class of dyes for cotton. These are applied in a soluble reduced form and after application they are oxidized, forming an insoluble molecule. Oxidation may be carried out by exposure to air, or more rapidly with hydrogen peroxide, sodium perborate, or potassium dichromate in dilute acetic acid. Azoic dyes are similar to the developed direct dyes except that the coupling compound is applied to the cotton first and the dye is then reacted with the coupling compound or developer. Sulfur dyes are similar to the vats in that they are applied in a soluble reduced form and subsequently oxidized on the fiber. The mordant dyes are used only to a limited extent on cotton and are most important for obtaining shades of black. Mineral dyes are formed by treating the cotton with a solution of a metallic salt and then precipitating the metal on the cotton. Iron and chromium salts are particularly useful for shades of olive, tan, and khaki.

Textile finishing with organic solvents is being seriously investigated, particularly dyeing with organic solvents(44). The use of chlorinated hydrocarbons instead of water seems to be the most economical solvent(45) so far. There are several solvent-dyeing techniques being evaluated and which becomes established depends on a variety of factors, since this new technology(46) requires tremendous changes involving new dyestuffs and machinery.

B. SOIL RELEASE

The term "soil release" in textiles can hardly be considered unless one includes a discussion of soil composition (see Table VI), soiling, and soil redeposition. Investigations have been carried out with a variety of soil compositions. In general, these soil formulations contain some fat component(s) and may or may not contain clays and carbon

TABLE VI
COMPONENTS INCLUDED IN SOIL COMPOSITIONS

(1) Fats
    (a) Fatty acids
    (b) Di- and triglycerides
    (c) Sterol esters
    (d) Paraffins, waxes, lubricating oils
(2) Clays
(3) Pigments, such as carbon blacks

blacks. Carbon blacks vary extensively in particle size and behavior, thus introducing additional control problems from one laboratory to another in preparing "soil compositions." Application of the soil to a woven fabric can, obviously, be influential in its removal. Excessive rubbing and working of the oily soil into the interstices and yarns in a fabric will permit more occlusion of the soil, particularly in fabric constructions which tighten up due to fiber swelling in laundering. Soiling and soil retention was described by Reeves and coworkers(47) as due to (1) interfacial attraction such as van der Waals forces, (2) electrostatic attraction, (3) mechanical forces, and (4) hydrophobicity and oleophilicity. In order to remove soil from a fiber surface by laundering, it is necessary for the free energy of the fiber–water interface to be lower than the free energy of the water–"soil" interface. This is aside from mechanical restraints due to yarn and fabric construction. Control of these interfacial energies are discussed in the literature in a variety of ways, such as "the fiber surface should be polar, hydrophilic, and negatively charged." Cotton fibers, after scouring and bleaching, present a polar and hydrophilic surface and the charge on the surface can be attained by the addition of anionic groups through the use of carboxymethyl cellulose(48).

This concern for soiling and soil release began when cotton fabrics were first treated with nitrogenous crosslinking agents for wash-and-wear fabrics. As the consumption of cotton–polyester blends increased and the fabrics were given durable-press treatments, serious problems arose in the areas of soiling, soil release (or retention), and redeposition on laundering. Solutions to this problem began with the incorporation of CMC in the crosslinking or resin-finishing bath. Other additives in resin finishing which have been evaluated include certain poly-acrylates(49) and a class of hybrid fluorochemical finishes(50). These polymeric chemicals impart to the fabric desirable hydrophilic, anionic, and oleophobic characteristics which improve antisoiling, soil release, and reduce soil redeposition in laundering.

C. FLAMEPROOFING

The development of chemical finishing procedures to impart flame retardance to textiles has become a major concern both to government and industry. The establishment of more stringent standards regarding the flammability of textiles may be anticipated through further federal legislation(51). In the case of cotton and other cellulosic fibers, flame retardance is achieved by inhibiting the formation of flammable volatile organic decomposition products and by catalyzing the dehydration reaction during the pyrolysis of cellulose.

Flame-retardant finishes for cotton may be bound to the fiber either chemically or physically. Most current processes have such drawbacks as causing loss of hand, stiffening the fabric, not inhibiting afterglow sufficiently, causing yellowing or other color changes upon laundering, reducing air permeability in the treated fabric, and reducing fabric strength, particularly if crosslinking is involved.

The pyrolysis of cellulose proceeds by a chain reaction mechanism (52) in which the initiation step involves scission of the glucosidic bond, and the major propagation step is the formation of levoglucosan. Cellulose exposed to high temperatures breaks down into a solid char and volatile liquids, including tars and bases. The tar is composed chiefly of levoglucosan with some carbonyl and unsaturated compounds. Flame-retardant finishes may operate by altering the course of the decomposition to produce a smaller amount of flammable volatile materials and a greater amount of solid carbonaceous char. This may occur by means of the flame retardant reacting with the cellulose at high temperatures to form intermediate compounds that decompose without forming flammable volatiles. The decomposition temperature of cellulose is lowered by all known flame retardants.

The carbonaceous char which is formed after the flaming of a treated cotton has a tendency to glow. This process, known as afterglow, is believed to be due to the following reactions for the oxidation of carbon:

$$(1) \quad C + \tfrac{1}{2}O_2 \longrightarrow CO \qquad H = 26.4 \text{ kcal/mole}$$
$$(2) \quad C + O_2 \longrightarrow CO_2 \qquad H = 94.4 \text{ kcal/mole}$$

To retard the afterglow, it is considered necessary to favor the first reaction by catalytic means in order to decrease the total heat produced. It has been shown that phosphorus-containing compounds are capable of reducing and even eliminating the afterglow.

Investigations of the kinetic and thermodynamic parameters for the pyrolysis of cellulose and treated celluloses have become more meaningful with the application of differential thermal analysis and thermal gravimetric analysis techniques(53). With these techniques, the energies involved in decomposition, as well as weight losses and residues, are obtained, thus allowing quantitative evaluation of the effects of flame retardants on the thermal decomposition process.

Although it is usually desirable to impart a durable flame- and afterglow-resistant finish, the compounds borax, ammonium sulfamate, ammonium sulfate, or phosphate can be used if there is no exposure to water. During World War II, a flame-retardant formulation used on fabrics for tents and tarpaulins consisted of antimony oxide, chlorinated paraffins, and a binder such as melamine or urea formaldehyde and calcium carbonate. A mildew inhibitor, such as copper naphthenate, and pigments were also included. Another finish similar to this is an aqueous dispersion of vinylidene chloride copolymer and antimony trioxide. It appears that certain metallic salts or oxides and chlorine-containing compounds are synergistic in imparting flame-retardant properties. Durability to laundering for a nonreactive retardant is obtained when tris(2,3-dibromopropyl) phosphate, and small quantities of polyvinyl chloride, polyvinyl acetate, or acrylic polymers are applied to fabrics in an organic solvent(54).

Durable flame-retardant systems include formulations based on tetrakis (hydroxymethyl) phosphonium chloride(55) (a trimethylolmelamine and urea) and tris(1-aziridinyl) phosphine oxide and thiourea. Tetrakis (hydroxymethyl) phosphonium hydroxide, urea, and trimethylolmelamine at a molar ratio of 2:4:1 in a pad-dry-cure process imparts flame resistance, good crease recovery angles, and minimal losses in breaking and tearing strength(56). Improved economics have been attained by Beninate(57) with a gaseous ammonia cure followed by a heat cure and the best nitrogen-to-phosphorus ratio was 0.9 to 2.1. Other durable flame retardants include dialkyl phosphono-carboxylic acid amides(58), n-methylol derivative of phosphonate-substituted amides(59), and cyanamide based phosphates(60).

The evaluation of fabric flammability by standard laboratory procedures poses many problems. It is necessary that these tests take into account the complete system and that consideration be given to such factors as the temperature and the rate of burning.

D. Rot Mildew and Light-Resistant Finishes

Although the requirements for rot resistance and light resistance are different, it seems appropriate to consider these finishes together. Cotton fabrics receiving these finishes are usually destined for outdoor applications in the form of tents, tarpaulins, awnings, and truck covers. Here they are exposed to the degrading effects of uv light, heat, microorganisms, and moisture.

The more significant species of fungi that are involved in the decomposition of exposed cotton fabrics include certain *Aspergillus, Chaetomium, Penicillium,* and *Fusarium,* while cellulolytic bacterial activity has been observed with *Sporocytophaga myxococcoides, Cellulomonas, Cellvibrio,* and *Cytophaga* species(*61*). These organisms do not always consume the cellulose itself but instead subsist on its decomposition products which are produced by the action of the enzymes secreted by the various organisms. These enzymes cause hydrolysis of the glucosidic linkages producing water soluble products. A study of slack mercerized and unmercerized cotton has indicated that microbial attack is not random, and the enzyme, being composed of large molecules with restricted mobility in the substrate, removes a number of adjacent glucose residues from its surroundings(*62*). Such enzymatic action may be associated with either a coenzyme or a portion of the surface of the enzyme. This portion or active site is responsible for the catalytic reaction. The coenzyme may be a smaller molecule which is associated in some way with the larger proteineous enzyme entity. Thus, to prevent this action, the active site may be blocked or the coenzyme may be reacted with some compound. Secondly, the cotton may be chemically modified to block the possible sites of enzymatic attack.

Many of the finishes used for protection from microorganisms are additive and may be classified as the phenolics, organometallics, quaternary ammonium compounds, and miscellaneous organic compounds, such as thiocarbamates.

1. Phenols

Among the phenol derivatives which have been found useful are pentachlorophenol, sodium and lauryl pentachlorophenol, pentachlorophenol-N-(bromophenyl) carbamate, and 2,2-dihydroxy-5,5-dichloro-diphenylmethane. Chlorinated compounds may, under certain conditions, decompose, with the possible formation of hydrochloric acid, which would be detrimental to the cotton.

2. *Organometallics*

These compounds compose the largest group of biocides which can protect cotton fibers. The metallic moiety includes copper, mercury, zinc, antimony, tin, bismuth, cobalt, boron, iron, zirconium, arsenic, and chromium. Copper 8-hydroxyquinolate is a much used compound in this group. Other examples are phenylmercuric acetate, tributyl tin oxide or its acetate, and triphenyltin chloride. These compounds have disadvantages of one type or another. For example, mercury derivatives are particularly toxic to humans, copper derivatives may cause odor, and tin derivatives are decomposed by sunlight.

3. *Miscellaneous Compounds*

Miscellaneous compounds include quaternary compounds such as perfluorinated derivatives of alkyl pyridines(*63*), sulfur compounds such as ditetrahydrothionaldehyde(*64*), and nitrogen compounds such as N-nitrophenyl- and N-nitrotolyl-itaconimide(*65*). There are also finishes based upon methylolmelamine which give durable protection from microorganisms by forming insoluble resins within the fiber when treated in the wet state(*66*). Cotton grafted with acrylonitrile at low add-on has also shown good resistance to microorganisms(*67*), as well as no loss in strength after 28 days exposure to spores of *Penicillim oxalicum*.

E. GRAFT POLYMERIZATION

Grafting of synthetic polymers to cotton is an effective method for modifying its properties. Grafting may be accomplished either by reacting a preformed polymer with the cotton or by actually causing polymerization to take place at active sites on the cotton cellulose. Reaction with preformed polymers are of limited use since diffusion of the high molecular weight synthetic polymer limits the grafting reaction to the fiber surface. The polymerization of a monomer to a cellulose chain can be accomplished by either ionic or free-radical initiation methods. The redox system and high-energy radiation techniques(*68*) have been extensively investigated. Avny and

Rebenfeld(69) have recently studied anionic graft polymerization using sodium cellulosates for initiation and suggested that the ionic technique provided better control of substitution on the hydroxyls and the molecular weight of the graft.

Some of the redox systems which have been studied for the graft polymerization of vinyl monomers onto cellulose include ceric salts (70), a ferrous salt–hydrogen peroxide(71), sodium thiosulfate-potassium persulfate(72), sodium periodate(73), manganic sulfate-sulfuric acid(74), and pentavalent vanadium(75). The vinyl monomers which have been grafted onto cotton fibers either by a redox system or by a radiation method include the following:

Acrylonitrile	Acrylates
	methyl
Methacrylates	ethyl
methyl	acrylamide
ethyl	methacrylamide
butyl	N-methylolacrylamide
hexyl	N,N-dimethylolacrylamide
lauryl	
stearyl	Styrene
benzyl	Vinyl acetate
tetrahydrofuryl	Vinyl chloride
glycidyl	Vinylidene chloride
Dimethacrylate	2-Vinylpyridine
ethylene	N-vinylpyrrolidone
propylene	Vinyl siloxanes
1,3-butylene	Vinyl phosphinic acid esters

Of the redox systems investigated, the tetravalent ceric ion has received considerable study because of its high grafting efficiency(76). When ceric ammonium sulfate is employed, 60°C is the preferred initiation temperature(77). Thus, the stability of the nitrate–ceric and sulfate–ceric complexes differ with the latter being more stable. Studies employing acrylonitrile as the monomer in such systems have produced polyacrylonitrile grafts with number-average molecular weights of approximately 50,000–55,000(78). As always expected in fiber research, the state of the fiber at the time of chemical reaction is an important factor in determining the diffusion and absorption of the monomer. Thus, the graft yields depend on fiber structure and micropore size, and such variables as temperature, concentration, nature of the vinyl monomer, and the reaction medium(79). Formation of homopolymer occurs in nearly all cases and excessive consumption of the initiator beyond that required by the cellulose can be used as a measure of the extent of homopolymerization. Extraction of homopolymer from the product with suitable solvents provides a more direct estimate of homopolymerization.

The site of the graft in the ceric ion redox system is still controversial. Some indications are that it may occur at the hemiacetal oxygen (the polymeric link), and also at adjacent hydroxyl units, possibly by oxidation of the OH groups(80). In this connection, Iwakura and coworkers have isolated from a cellulose-styrene graft copolymer both block and graft copolymers. They have further verified these reactions in a model system using various alcohols. Grafting on cellulose is also affected by the presence of carbonyl and aldehyde groups, particularly at low concentrations of the ceric salt(81).

There are three principal methods of radiation-induced grafting: preirradiation, simultaneous irradiation, and postirradiation(82). The preirradiation method involves exposure of the polymeric substrate to the radiation in order to produce free radicals and the subsequent application of the liquid or gaseous monomer. The simultaneous technique involves the radiation of the polymeric substrate in the presence of the monomer. A major drawback to this technique is that it usually leads to the formation of excessive homopolymer. Postirradiation involves initially the application of monomer followed by exposure to an energy source.

Water and other swelling agents have an important effect on the extent of graft polymerization and on the location of the graft polymer in a cotton fiber(83). This was first demonstrated by Stannett and Kesting who investigated the effect of water in a styrene–dioxane solution used for preirradiation grafting to cotton. Significant polymer add-on was not observed until the water content in the monomer solution rose above a certain minimum amount(84). The fact that diffusion of the monomer, acrylonitrile, controls the final distribution of the polyacrylonitrile grafts within the fiber is also well demonstrated by the photomicrographic technique of Rollins and coworkers(85). Nonuniform and surface grafting occurs when the solvent has no swelling action on the cellulose and when the fiber structure has not been opened by a chemical pretreatment, such as mercerization or derivative formation. In radiation grafting with higher doses or with a swollen substrate, shorter graft chains can usually be obtained; however, the effects of dose strengths on the reaction yields, the length of the graft chains, and their distribution in the fiber cross section depends on the specific method used(86). It should also be noted that irradiation of a cotton fiber results in chemical degradation and changes in the physical structure, the extent of which is influenced by environmental conditions and energy dosage. It is possible to control process conditions so that excessive strength losses are prevented.

Graft polymers on cotton may be used to waterproof, improve rot resistance, flameproof, change dye characteristics, and modify physical properties, such as abrasion resistance and stiffness. For example, propyl or butyl acrylate-grafted cotton is claimed to have remarkably improved physical properties (87). Although not a graft, the formation and deposition of elastomers from butadiene and isoprene inside cotton fibers gives improved resilience (88).

VI.   Other Cellulose Fibers

In addition to cotton there are many natural cellulosic fibers in commercial use today. This important group of fibers includes, principally, flax, ramie, hemp, jute, and sisal (89). Sisal is a leaf fiber while the others are vegetable fibers which are located in the stems of plants and are referred to as bast fibers. These fibers contain approximately 70–76% cellulose, with varying quantities of hemicelluloses, and pectius lignin. In preparation for commercial use these fibers are subjected to a number of chemical purification treatments after their isolation from the plant. The removal of hemicellulose and lignin from flax is accomplished by a chemical or microbial treatment to yield the flax fiber which is referred to as linen. The microbial treatment known as retting in water is best, but depending on the method, can produce fibers varying in color from cream to a dark brown. Hemp and jute fibers are also obtained from a similar retting process but ramie, found in the bark of the stem, cannot be extracted in this manner. To isolate ramie fibers, a mechanical scraping of the bark is required in order to remove extraneous matter. Then a degumming process which includes both chemical and mechanical techniques is used for further purification.

Flax and ramie fibers are superior in strength and versatility to other bast fibers, and each can be spun, knitted, and woven by itself or in blends with other fibers. The difficulties in processing and degumming ramie have kept it from being commercially competitive with flax.

REFERENCES

1. W. A. Sisson and G. L. Clark, *Ind. Eng. Chem.. Anal. Ed.*, **5**, 296 (1933).
2. M. L. Rollins and V. W. Tripp, *Text. Res. J.*, **24**, 345 (1954).

3. S. Hanley, ed., "The American Cotton Handbook," Wiley-Interscience, New York, 1965, Chpt. 3 by Mary L. Rollins.
4. G. Raes, T. Fransen, and L. Verschraege, *Text. Res. J.*, **38**, 182 (1968).
5. L. Segal, *Text. Res. J.*, **24**, 861 (1954).
6. L. Segal, *J. Poly. Sci.*, **55**, 395 (1961).
7. L. Loeb and L. Segal, *J. Poly. Sci.*, **14**, 121 (1954).
8. C. Y. Liang and R. H. Marchessault, *J. Poly. Sci.*, **37**, 385 (1959); *ibid.*, **39**, 269 (1959); *ibid.*, **43**, 71 (1960).
9. L. Rebenfeld and W. P. Virgin, *Text. Res. J.*, **27**, 286 (1957); L. Rebenfeld, *Text. Res. J.*, **31**, 123 (1961); *ibid.*, **32**, 154 (1962).
10. E. R. Trotman, "Textile Scouring and Bleaching," Charles Griffin and Co., London, 1968, p. 18.
11. P. W. Heterington, *J. Soc. Dyers Colour.*, **84** (7), 359 (1968).
12. W. K. Walsh, C. R. Jin, and A. A. Armstrong, *Text. Res. J.*, **35**, 648 (1965).
13. P. K. Shenoy and J. W. Pearce, *Am. Dyest. Rep.*, **57**, 352 (1968).
14. U.S. Patent 2,974,432, Koret of California.
15. U.S. Patent 2,957,746, National Cotton Council of America.
16. U.S. Patent 2,950,553, Rohm and Haas Co.
17. J. Hepp, *Chemiefasern.* **18**, 188, 190 (1968).
18. E. J. Blanchard, R. J. Harper, Jr., G. A. Gautreaux, and J. David Reid, *Amer. Dyest. Rep.*, **57**, 610 (1968).
19. V. W. Tripp, A. T. Moore, I. V. deGruy, and M. L. Rollins, *Text. Res. J.*, **30**, 140 (1960).
20. V. W. Tripp, A. T. Moore, and M. L. Rollins, *Text. Res. J.*, **31**, 295 (1961).
21. M. L. Rollins, A. T. Moore, and V. W. Tripp, *Text. Res. J.*, **33**, 117 (1963); J. G. Frick, Jr., *et al.*, *Amer. Dyest. Rep.*, **52**, 953 (1963).
22. S. P. Rowland and A. W. Post, *J. Appl. Poly. Sci.*, **10**, 1751, (1966).
23. S. P. Rowland, M. L. Rollins, and I. V. deGruy, *J. Appl. Poly. Sci.*, **10**, 1763 (1966).
24. S. M. Stark, Jr., and S. P. Rowland, *J. Appl. Poly. Sci.*, **10**, 1777 (1966).
25. G. K. Joarder, and S. P. Rowland, *Text. Res. J.*, **37**, 1083 (1967).
26. R. Jeffries, J. G. Roberts, and R. N. Robinson, *Text. Res. J.*, **38**, 234 (1968).
27. S. P. Rowland, A. L. Bullock, V. O. Cirino, E. J. Roberts, D. E. Hoiness, C. P. Wade, M. A. F. Brannan, H. J. Janssen, and P. F. Pittman, *Text. Res. J.*, **37**, 1020 (1967).
28. I. Croon, *Svensk Papperstidn.*, **63**, 247 (1960).
29. E. J. Roberts and S. P. Rowland, *Carbohydrate Res.*, **5**, 1 (1967).
30. S. P. Rowland, A. L. Bullock, V. O. Cirino, and C. P. Wade, *Can. J. Chem.*, **46**, 451 (1968).
31. S. Haworth, J. G. Roberts, and R. N. Robinson, *Textilveredlung*, **2**, 361 (1967).
32. C. P. Wade, E. J. Roberts, S. P. Rowland, *J. Poly. Sci.*, *Pt. B*, **6**, 673 (1968).
33. J. J. Willard, R. Turner, and R. F. Schwenker, Jr., *Text. Res. J.*, **35**, 564 (1965).
34. S. Patel, Joseph Rivilin, T. Samuelson, O. A. Stamm, and H. Zollinger, *Text. Res. J.*, **38**, 226 (1968).
35. H. M. Ziifle, R. J. Berni, and R. R. Benerito, *Text. Res. J.*, **31**, 349 (1961); H. M. Ziifle, R. J. Berni, and R. R. Benerito, *J. Appl. Poly. Sci.*, **7**, 1041 (1963).
36. E. J. Gonzales and R. R. Benerito, *Text. Res. J.*, **35**, 168 (1965); E. J. Gonzales, R. R. Benerito, and R. J. Berni, *Text. Res. J.*, **36**, 565 (1966).
37. E. J. Gonzales, R. R. Benerito, R. J. Berni, and H. M. Zacharis, *Text. Res. J.*, **36**, 571 (1966).
38. H. M. Ziifle, R. R. Benerito, E. J. Gonzales, and R. J. Berni, *Text. Res. J.*, **38**, 925 (1968).

39. E. J. Gonzales, H. M. Ziifle, R. J. Berni, and R. R. Benerito, *Text. Res. J.*, **37**, 726 (1967).

40. R. J. Berni, R. M. Reinhardt, and R. R. Benerito, *Text. Res. J.*, **38**, 1072 (1968).

41. M. S. Haith and C. G. C. Namboori, *Text. Res. J.*, **38**, 1061 (1968).

42. R. B. Penland, S. Mizushima, C. Curran, and J. V. Quagliano, *J. Amer. Chem. Soc.*, **79**, 1575 (1957).

43. I. D. Rattee, *J. Soc. Dyers Colour.*, **85**, 23 (1969).

44. Anonym., *Text. Ind.*, **130**, 291 (1966).

45. G. Siegrist and J. R. Geigy, *Textilveredlung*, **4** (1), 12 (1969).

46. B. Milićević, *Textilveredlung*, **4** (4), 213 (1969).

47. W. A. Reeves, J. V. Beninate, R. M. Perkins, and G. L. Drake, Jr., *Amer. Dyest. Rep.*, **57**, P1053 (1968).

48. J. V. Beninate, E. L. Kelly, G. L. Drake, Jr., and W. A. Reeves, *Amer. Dyest. Rep.*, **55** (2), 25 (1966).

49. A. Feinauer, H. Billie, and W. Ruttiger, *Amer. Dyest. Rep.*, **58** (6), 16 (1969); H. E. Billie, A. Eckell, and G. A. Schmidt, *Text. Chem. Color.*, **1**, 600/23 (1969).

50. S. Smith, P. O. Sherman, and B. Johannessen, *Text. Res. J.*, **39**, 441, 449 (1969).

51. Chem. and Eng. News, April 8, 1968, p. 23.

52. P. K. Chatterjee and C. M. Conard, *Text. Res. J.*, **36**, 487 (1966).

53. K. Akita and M. Kase, *J. Poly. Sci.*, **A5**, 833 (1967).

54. T. D. Miles and A. C. Delasanta, *Text. Res. J.*, **38**, 273 (1968).

55. U.S. Patent 3,404,022, 3,403,044, U.S. Government.

56. J. V. Beninate, E. K. Boylston, G. L. Drake, Jr., and W. A. Reeves, *Text. Res. J.*, **38**, 267 (1968).

57. J. V. Beninate, 17th Chemical Finishing Conf., National Cotton Council, Washington, D.C., October 1968.

58. R. Aenishanslin, 17th Chemical Finishing Conf., National Cotton Council, Washington, D.C., October, 1968.

59. G. C. Tesoro, S. B. Sello, and J. J. Willard, *Text. Res. J.*, **38**, 245 (1968).

60. S. J. O'Brien, *Text. Res. J.*, **38**, 256 (1968); British Patent 1,110,116, Courtaulds.

61. R. G. H. Siu, "Microbial Decomposition of Cellulose," Reinhold, New York, Part C.

62. K. Selby, *Biochem. J.*, **79**, 562 (1961).

63. U.S. Patent 3,147,064, Minnesota Mining and Mfg. Co.

64. British Patent 932,152, DuPont Co.

65. British Patent 930,450; 930,598, U.S. Rubber Co.

66. *Ciba Rev.* (6), 38 (1961).

67. A. A. Armstrong and H. A. Rutherford, *Text. Res. J.*, **33**, 264 (1963).

68. H. A. Krässig and V. Stannett, *Adv. Poly. Sci.*, **4**, 111 (1965).

69. Y. Avny and L. Rebenfeld, *Text. Res. J.*, **38**, 559, 684 (1968).

70. A. Y. Kulkarni and P. C. Mehta, *J. Poly. Sci.*, **9**, 2633 (1965); Y. Iwakura, T. Kurosaki, and Y. Imai, *J. Poly. Sci.*, **A3**, 1185 (1965).

71. J. C. Arthur, Jr., O. Hinojosa, and M. S. Bains, *J. Appl. Poly. Sci.*, **12**, 1411 (1968); Yoshitaka Ogiwara, Yukie Ogiwara, and Hitoshi Kubota, *J. Appl. Poly. Sci.*, **12**, 2575 (1968).

72. A. Y. Kulkarni, A. G. Chitale, B. K. Vaidya, and P. C. Mehta, *J. Appl. Poly. Sci.*, **7**, 1581 (1963).

73. T. Toda, *J. Poly. Sci.*, **58**, 411 (1962).

74. H. Singh, R. T. Thompy, and V. B. Chipalkatti, *J. Poly. Sci.*, **A3**, 4289 (1965).

75. R. M. Livshits, L. M. Livites and Z. A. Rogovin, *Vysokomolekul. Soedin*, **6**, 1624 (1964).

76. S. Kaizerman, G. Mino, and L. F. Meinhold, *Text. Res. J.*, **32**, 136 (1962).

77. A. Hebeish and P. C. Mehta, *Text. Res. J.*, **37**, 911 (1967); A. Hebeish and P. C. Mehta, *J. Appl. Poly. Sci.*, **12**, 1625 (1968).

78. A. Y. Kulkarni and P. C. Mehta, *J. Appl. Poly. Sci.*, **12**, 1321 (1968).

79. R. M. Huang and P. Chandramouli, *J. Appl. Poly. Sci.*, **12**, 2549 (1968).

80. Yoshio Iwakura, Toshikazu Kurosaki, and Yohji Imai, *J. Poly. Sci.*, **A3**, 1185 (1965).

81. Yoshitaka Ogiwara, Yukie Ogiwara, and Hitoshi Kubota, *J. Poly. Sci.*, **A5**, 2791 (1967).

82. V. Stannett, "Proceedings International Symp. Radiation-Induced Polym. and Graft Copolymer," Battelle Memorial Institute, November 1962, p. 259.

83. F. L. Saunders and R. C. Sovish, *J. Appl. Poly. Sci.*, **7**, 357 (1963).

84. R. E. Kesting and V. Stannett, *Makromol. Chem.*, **55**, 1 (1962).

85. M. L. Rollins, A. M. Cannizzaro, F. A. Blouin, and J. C. Arthur, *J. Appl. Poly. Sci.*, **12**, 71 (1968).

86. L. Wiesner, *Melliand Textilber.*, **49** (1), 99 (1968).

87. Michiharu Negishi, Yoshio Nakamura, Toshiko Kakinuma, and Yoriko Iizuka, *J. Appl. Poly. Sci.*, **9**, 2227 (1965).

88. D. Meimoun and A. Parisot, *Text. Res. J.*, **39**, 560 (1969).

89. R. H. Kirby, "Vegetable Fibers," Wiley-Interscience, New York, 1963.

Chapter 2 RAYON

I. Introduction

The cellulosic fiber rayon was the first commercial manmade fiber, and today it is still the most important. The most commonly used process by which this fiber is made, the viscose process, is founded on the work of Cross and Bevan (1891–1892). Pioneering of this process was carried out by Courtaulds, Ltd., of England and involved the solubilization of a pure cellulose by reacting it with carbon disulfide in the presence of sodium hydroxide. This viscous solution, called viscose or viscose dope, is extruded through spinnerets into an aqueous acid bath which coagulates the cellulose xanthate, decomposes it, and regenerates the cellulose. The viscose process of today utilizes the same chemical reactions, but there are many variations and additions to the process. These improvements have led to rayon fibers with not only a variety of physical properties but with physical properties which have made it highly competitive with other cellulosic fibers, such as cotton and flax.

Another method for the manufacture of rayon is called the cuprammonium process. This process originally utilized cotton linters as the source of purified cellulose, but a high α-cellulose wood pulp may also be used. The cellulose is solubilized by treating it in ammonia and basic copper sulfate and subsequently a sodium hydroxide solution. This mass is kneaded until a clear blue "solution" is obtained because of the ammonical copper complex formed with the cellulose. The solution is diluted to give a 9–10% cellulose content, filtered, deaerated, and extruded through a spinneret into a water coagulating bath. The fibers then must be given extensive washings to remove all the salts. This rayon is called cuprammonium rayon and is still produced in limited quantities for a variety of specialty yarns, particularly of the slubbed type.

A third rayon fiber, produced commercially under the trade name Fortisan, is made by spinning the fiber as a cellulose acetate. This fiber is subjected to high stretching under steam and then saponified to remove the acetate groups, leaving a regenerated cellulosic fiber.

There continues to be an active search for a solvent for cellulose that would replace the derivatization and subsequent destruction of the cellulose derivative in the spinning process. The Jayme-type solvents were encouraging, but adequate concentrations of cellulose for spinning purposes could not be achieved. Cellulose dissolved in a "mixture" of dimethyl sulfoxide, nitrogen dioxide, and a trace quantity of water can be extruded as fibers and films(*1*). Another recently reported solvent for cellulose is an amine liquid sulfur dioxide solution(*2*). The search for suitable cellulose solvents continues and no doubt other processes for preparing rayon fibers will emerge.

II. Fiber Formation

A. Viscose Process Chemistry

Purified bleached wood pulp is the source of cellulose used in the viscose process. This pulp, in the form of sheets, is soaked in 17.5% sodium hydroxide for 1–4 h and then pressed to remove the excess alkali containing alkali-soluble hemicelluloses. The pressed material, called alkali or soda cellulose, is shredded until it forms fine crumbs. The crumbs are then aged in the presence of oxygen or air during which the degree of polymerization is reduced by a random oxidative cleavage of the cellulose chains. After ageing, the alkali cellulose is reacted with carbon disulfide in a rotating air-tight drum to form cellulose xanthate. The crumb develops a deep orange color which is attributed to the formation of trithiocarbonate, carbonate, sulfides, and polysulfides. The main reactions are usually written as follows:

$$\text{Cell-OH} + \text{NaOH} \longrightarrow \begin{bmatrix} \text{Cell-ONa} \\ \text{Cell-OH} \cdot \text{NaOH} \end{bmatrix}$$
$$\text{alkali cellulose}$$

$$\text{Cell-ONa} + \text{CS}_2 \longrightarrow \text{Cell-OC-SNa}$$

sodium cellulose xanthate

It has not been firmly established whether a true sodium salt of cellulose (sodium cellulosate above) is formed or whether the product should be viewed as a hydrated adduct between NaOH and cellulose, Cell-OH·NaOH. Under any circumstance, alkali cellulose is highly reactive and behaves as a sodium salt of cellulose (sodium cellulosate) in many reactions.

Commercially, the quantity of carbon disulfide added is about 70% of the theoretical quantity required for a DS (degree of substitution) of one. By increasing the carbon disulfide and reaction time, it is possible to increase the degree of xanthation.

After solution of the cellulose xanthate in sodium hydroxide, and the addition of additives if desired, the so-called viscose is deaerated and allowed to "ripen." Ripening is basically storage at controlled temperatures. During this time, changes in the distribution of the xanthate groups occur both within a cellulose chain as well as among cellulose chains. These changes may be simply described as follows:

Xanthate decomposition

$$\text{ROCS}_2^- \longrightarrow \text{RO}^- + \text{CS}_2$$

Rexanthation

$$\text{R'O}^- + \text{CS}_2 \longrightarrow \text{R'OCS}_2^-$$

Byproduct formation

$$2\text{CS}_2 + 6\text{NaOH} \longrightarrow \text{Na}_2\text{CS}_3 + \text{Na}_2\text{CO}_3 + \text{Na}_2\text{S} + 3\text{H}_2\text{O}$$

Xanthate group redistribution

$$\text{ROCS}_2 + \text{R'OH} \rightleftharpoons \text{R'OCS}_2^- + \text{ROH}$$

The rates and extent of these changes are followed by two measurements, the ball-fall viscosity, and the salt test (a measure of coagulability). Ball-fall viscosity passes through a minimum as a function of time and then increases to a gel point. In the salt test, the coagulability of the viscose is related to the degree of substitution by xanthate groups and it decreases as xanthate groups are lost by decomposition.

The spinning of viscose into a fiber involves the basic reaction of regeneration and is simply the recovery of cellulose from cellulose xanthate by an acid decomposition reaction. Upon spinning into the

$$ROCS_2^- + H^+ \longrightarrow ROH + CS_2$$

acid bath, the viscose is coagulated and the acid diffuses into the fiber regenerating the cellulose. During this time, the fibers are also stretched and the cellulose molecules are oriented in the direction of the fiber axis.

The variety of decomposition reactions for cellulose xanthate have been extensively studied by Phifer and Dyer(3). Figure 1 shows their proposed decomposition pathways which ultimately lead to the products shown above.

The spinning bath systems used today may be described as (1) the zinc-based spin bath and (2) the non-zinc spin bath. Systems which use zinc are employed in the production of regular and crimped rayons, rayon tire yarn, and modified high wet modulus (HWM) rayons. The non-zinc spin bath is used in the production of polynosics. Over 95% of the rayon produced today is spun with zinc in the spin bath. This bath is aqueous and contains sulfuric acid, sodium sulfate, and zinc sulfate(4). The nonzinc process includes aqueous spin baths containing (1) sulfuric acid and sodium sulfate (producing normal rayon, Toramomen, or polynosic type), (2) sulfuric acid (Lilienfeld), and (3) phosphate, pH 4 (Fiber G). Formaldehyde may be a bath additive for some polynosic spinning processes.

B. Structure of Rayons

It has been estimated that there are more than 64 independent variables in the production of viscose rayon. The structure of the fiber and the resulting physical and chemical properties are controlled by these variables, some of which are listed as follows:

Unremoved α-cellulose pulp impurities
Degree of polymerization of cellulose (and DP distribution)
Degree of xanthation
Xanthation by-products
Quantity of cellulose in the viscose
Alkali concentration
Modifiers
Filterability
Deaeration

Spinning Variables
    Bath composition
    Concentration of bath components
    Bath additives (usually organic compounds)
    Temperatures
        (a) coagulation and/or regeneration baths
        (b) washes on godets
        (c) a second bath
    Pump speed (into bath)
    Stretch — location and amount
    Extent of removal of all impurities in fiber
    Drying conditions

These are only a few of the variables that are involved. The degrees of structural variations which are achieved through control of these variables lead to rayons of quite different properties. Daul and Muller (5) have divided rayon fibers into five types: (1) regular; (2) low wet modulus, high tenacity; (3) intermediate wet modulus, high tenacity; (4) high wet modulus, high tenacity; and (5) polynosic (also a high wet-modulus type).

Table I outlines the general characteristics of these fiber types. Perhaps it might be easier to discuss the fiber types as if there were only three general structures: (1) polynosic, (2) high wet modulus, and (3) regular. The polynosic fiber is formed under such conditions that the cellulosic crystal nuclei, which are relatively few in number, have a chance to grow into well-developed long crystallites. The unregenerated coagulated filaments are highly stretched to give a fiber with a very high orientation in its structure upon regeneration. Such a treatment yields a fiber containing a network of long, crystalline fibrils having noncrystalline regions which are partially oriented and somewhat inaccessible. High wet-modulus rayons are made from cellulose of a lower degree of polymerization than that used for polynosics (see Table I). The method of production yields a highly oriented fiber also, but the length and thickness of the crystallites are somewhat less than those in the polynosics (6). Regular rayons contain even lower DP cellulose, have smaller crystallites and an overall lower orientation and degree of lateral order.

The recognition that the presence of zinc ions in the spin bath slowed down the regeneration process by forming zinc cellulose xanthate, which is more resistant to acid decomposition (7), began the development of the high-tenacity rayons of today. The role of the zinc is, in effect, to crosslink the cellulose chains in the form of zinc cellulose xanthate. The coagulated "crosslinked" fiber is then capable of more stretching (and apparently molecular orientation) than the

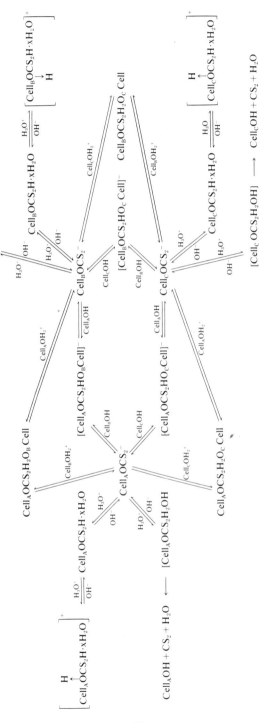

Fig. 1. Decomposition scheme for cellulose xanthate (FMC Corp.).

TABLE I
CHARACTERISTICS OF VARIOUS RAYON FIBERS(5)

	Regular	LWM(HT)*	IWM(HT)*	HWM(HT)*	Polynosic
Tenacity, g/den					
conditioned	1.5–2.8	3.0–5.0	3.5–5.0	4.0–7.5	3.6–5.0
wet	0.7–1.6	1.9–3.8	3.0–3.8	3.0–6.0	2.5–4.0
Elongation, %					
conditioned	15–30	14–25	12–17	5–10	7–11
wet	20–35	17–30	14–20	6–11	8–15
Wet modulus, g/den.	0.12–0.25	0.12–0.25	0.5–0.7	1.8–2.5	0.9–1.1
Loop strength, g/den	1.0–1.5	1.0–2.5	0.6–1.1	0.7–0.95	0.5–0.9
Knot strength, g/den.	0.7–1.6	1.7–2.5	0.6–1.8	1.2–2.8	1.2–2.9
Tensile strength, 1000 psi	29–46	58–90.5	60–100	80–156	60.5–120
Moisture regain, %	12–15	12.5–14.5	11–14	10–12	10–12
Fiber DP	300	300–500	350–600	600–800	550–650
Modifier	none	PEG + amine	PEG + amine	PEG, amine, HCHO	none
Stretch, %	60–120	100–120	140–150	200–400	200–300
Crystal size	small	small	intermediate	large	large
Cross section	skin core	all skin	all skin	layered	all skin
Dimensional stability	low	low	good	excellent	excellent
Caustic resistance	low	low	good	excellent	excellent
Application	versatile	industrial, blends	blends	100%, blends	100%, blends

*L is low; I, intermediate; and H, high wet-modulus(WM) rayons of high tenacity(HT).

uncrosslinked sodium counterpart. However, Dyer and Phifer(8) find crosslinking extremely unlikely and postulate that the most probable form of the zinc derivative is

$$\text{Cell OS} \overset{\displaystyle S}{\underset{\displaystyle S\text{-}Zn^+\,HSO_4^+}{\diagup\diagdown}} \qquad \text{or Cell OC} \overset{\displaystyle S}{\underset{\displaystyle S\text{-}Zn^+\,HCS_3^-}{\diagup\diagdown}}$$

Viscose regeneration modifiers have aided in the development of these new rayons. The compounds currently used include: aliphatic amines, ethoxylated amines, polyethylene glycols, ethoxylated phenol, aromatic amines, and quaternary ammonium salts. The viscose modifiers slow down the regeneration reaction, although their exact mode of action is not clear. It may be that they form a thin film on the fiber surface, thereby slowing down acid diffusion(9), or if an amine, forming insoluble zinc dithiocarbamate which acts as a barrier to diffusion (10). Many suggestions have been offered to explain the role of the viscose modifiers. Whether they act by forming diffusion controlling products or more acid-stable xanthates, they allow considerable variations in the viscose process and a corresponding variation in rayon fiber properties.

In addition to the broad types of rayons described in Table I, there are certain specialty rayons which are in limited commercial production. These include (1) multicellular rayons which may be flat or tubular in cross section(11), (2) rayons of high liquid absorbency(12), (3) rayons containing inorganic compounds which upon firing yield ceramic fibers(13), and (4) wool-like rayons of exceptionally high crimp and resiliency.

The usual analytical tools of x-ray diffraction, electron microscopy, birefringency, staining techniques, and absorption methods have been used in characterizing rayon structures. For future development, it would seem that ionic etching of the fiber surface(14) in the presence of helium, argon, nitrogen, or oxygen with subsequent study of the surface by electron microscopy would reveal additional structural knowledge. Also, the application of broadline nuclear magnetic resonance to structural studies, particularly in the presence of water and deuterium oxide, might prove to be of value(15).

Regular rayon has a skin-core structure in its cross section as observed by a differential staining technique(16). The skin is composed of many small crystallites as contrasted to the core of the fiber which is composed of a coarser crystalline structure in an amorphous matrix. The shape of the cross section of rayons can vary from a serrated irregular structure (regular), through a rectangular or branched outline, to an ovular and circular shape (HWM). Most rayons show some skin-core structure, with the exception of the polynosics which are all core and the tire yarns which are all skin.

The cross sections of all rayon fibers contain voids of less than 0.5 μ, which Sisson(17) has attributed to the formation of gas bubbles. The shape, size, and number of voids vary from one rayon to another. In a study of rayon tire yarns, Kaeppner(18) found that the smaller the voids the better the wet strength, and conversely the shorter the voids the better the flex life. He also observed that with an increasing ratio of void length to diameter, the flex life is shortened. Thus, a decrease in flex life caused by stretching may be attributed to an increase in orientation and/or an increase in void-length–diameter ratio. The high wet-modulus rayons have very small voids(19) as well. This may be related to the effectiveness of the viscose additives in the dispersion of gas bubbles. A further contribution to the study of voids was made by Gröbe and Gensrich(20). Their electron microscope observations of ultrathin cross sections of coagulated but unregenerated filaments showed no voids, but rather the fine structure of a cellulose network.

III. Properties of Rayons

A. PHYSICAL

Many variations in the physical properties of rayon fibers are possible because the manufacturer has the ability to control both the chemical reactions and the physical changes which take place during the spinning process. These property variations are illustrated in Table I. The importance of the new family of fibers, high wet-modulus rayons, is still growing, due principally to their wet stress–strain behavior which is similar to cotton. In addition to other structural characteristics which have been discussed, the HWM rayons have smaller micropores, less internal surface area, and a well-ordered and hence denser solid state. Most rayons, regardless of type, have about 40% crystallinity. In order to appreciate the versatility of the rayon fibers, careful consideration should be given to the stress–strain curves shown in Figs. 2a and 2b(21).

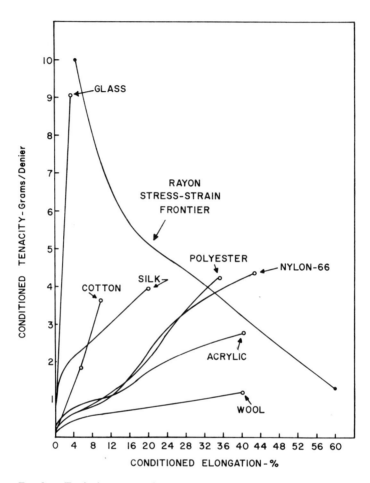

FIG. 2a.   Typical stress–strain curves for natural and synthetic fibers (21).

B. CHEMICAL

The chemical properties of rayons are very similar to those of cotton, since both fibers are cellulosic in nature. Their resistance to alkali ranges from poor to good depending on physical structure. Rayon hydrolyzes and disintegrates in mineral acids and, in the presence of oxidizing agents, strength losses may be incurred with increases in carbonyl and carboxyl groups. The rayons are little affected by the common bleaching agents. High wet-modulus rayons require no special precautions when bleached with sodium chlorite or hypochlorite.

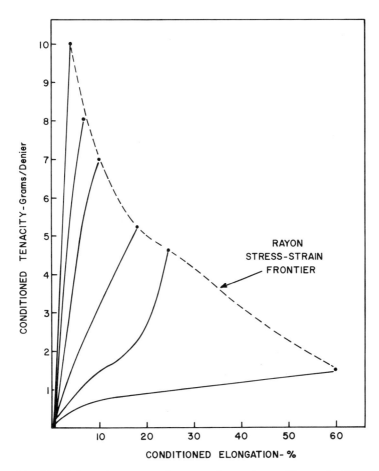

FIG. 2b. Schematic stress–strain curves of several rayon fibers today (*21*).

If bleaching is carried out with hydrogen peroxide, however, it is necessary to include a sequestrant and to control the temperature carefully (*22*). If the fabric is a polyester–HWM blend, sodium chlorite is best for bleaching. For good alkali resistance during mercerization in a 50–50 blend with cotton, the washing operation should be rapid, for even HWM rayon will show some degradation as the sodium hydroxide concentration passes through the 8–10% range. Polynosic behavior in sodium hydroxide might best be described as similar to regular rayon. In 5% NaOH at 20°C, shrinkage was 14.9% and weight loss was approximately 10% (*23*). Rayons are soluble without degradation in such well-known systems as aqueous solutions of quaternary ammonium compounds at 0°C and metal complex solutions such as cupriethylene diamine (Cuene) and cadmium ethylene diamine (Cadoxene).

IV.   Fiber Modifications

A. CROSSLINKING

Many of the crosslinking agents, thermosetting resins, and vinyl monomers that have been applied to cotton have also been evaluated on rayon fibers and fabrics. The general conclusion reached several years ago was that about twice the quantity of reagent was required on rayon to achieve the same level of smooth drying and crease resistance as was required for cotton. In view of the fact that this conclusion refers to regular rayon, a brief historical review of rayon in the typical pad-dry-cure processes should be considered.

In 1954, Cooke, Dusenbury, Kienle, and Lineken(24) showed that wrinkle resistance of a fabric is translatable through the yarns to the "elastic" properties of the fibers and that the ability to recover from creasing is more important than the ability to resist creasing. When a regular rayon is crosslinked, the change in crease recovery with water correlates well with the variation in fiber resiliency versus water content(25). (At a 0% r.h., even an uncrosslinked rayon shows a high crease recovery.) As Morton and Beaumont(26) showed earlier, the properties of elongation, dry elastic recovery, wet initial modulus, and water imbibition are interrelated in a way that "does not depend materially on the type of finishing treatment used."

For any cellulosic structure, a rayon or cotton, there is a compromise, within certain limits for each structure, between the amount of improvement desired in such properties as elastic recovery, imbibition, and wet modulus, and the amount of reduction which can be tolerated in such properties as tensile strength, extensibility, and abrasion resistance. This does not include the modest improvements which can be brought about through fabric construction and the use of lubricants and softeners. The attainment of low levels of water imbibition may be one of the most desirable properties to achieve in a rayon fiber, whether it is a result of the fine structure created during the spinning process, the formation of crosslinks in the cellulose, by resin deposition in fiber, the grafting of vinyl monomers, or any combinations(s) of these. Robinson (27) has stated that the most useful crease resistant finish for rayon fabrics is found in the urea–formaldehyde and melamine–formaldehyde precondensates. These are products which usually react with one another and simultaneously, but to a lesser extent, with the hydroxyl groups of cellulose to form crosslinks. The more "reactive resins" (DMEU, DMDHEU) react with one another less rapidly and with the cellulosic hydroxyl groups more readily, and seem to be particularly suitable for cotton fabrics.

Another approach to improving the crease recovery of a rayon fabric was undertaken by fiber manufacturers and this was to crosslink and resinify the fiber in bulk form(28). This was accomplished with a formulation utilizing urea, melamine, epichlorohydrin, and formaldehyde(29) and the objective was a crosslinked rayon processable on the cotton spinning system. Rayon manufacturers will remember these fibers although they are no longer produced in the U.S. Such a fiber would still be desirable provided its stiffness could be reduced.

Single fiber studies of a regular rayon have been carried out by Rebenfeld(30) using the crosslinking agent dimethylol ethylene urea. The fibers were reacted under tensions from 0 to 420 mg. When crosslinked under tension they were stronger, more resilient, stiffer, and less extensible than when crosslinked under zero tension. However, compared to a similar treatment of single cotton fibers, the effects of the tension on rayon were not as great. Unfortunately, due to differences in fiber diameters, the tensile loads corresponded to stresses of 0.4 mg/cm^2 in the case of cotton and only 0.22 mg/cm^2 for rayon. Since a regular rayon is a structure of low order, Rebenfeld suggests that a higher axial tension might be more desirable.

The high wet-modulus rayons are fibers of higher order with physical properties approaching those of cotton, and their requirements in a resin-finishing process are thus different from a regular rayon. These fibers require less crosslinking reactant than regular rayon because they already have improved wet and dry physical properties. When crosslinking a high wet-modulus rayon, it has been found that less dimethylol ethylene urea (DMEU) is required than a urea–formaldehyde compound(31). All of these compounds perform equally well in a pad-dry-cure-type treatment. However, the urea–formaldehyde- and triazone-based resins appear to be better in that they give good wrinkle recovery with less strength losses. If these rayons are finished for minimum care performance in blends other than cotton, better results are obtained if the reactant is increased 10–20%. The rayon which appears to be best in durable-press garments is a fiber described as an intermediate–high wet-modulus, high-tenacity fiber(32).

In durable-press garments of cotton or rayon, or blends containing these, it is possible to remove the "permanent" creases and form new "permanent" creases(*33*). This can be accomplished by treating the crease areas with an aqueous solution of an organic acid, such as citric acid or citric acid mixed with a metal salt such as magnesium chloride, and pressing the garment at temperatures over 200°F.

The best rayon fiber structure for finishing with an easy care resin treatment or a durable-press treatment has really not been determined. The literature offers no significant information on this nor does there appear to have been any careful systematic study on the subject. Of particular interest would be the abrasion resistance after resin treatment of fibers with a variety of fine structures.

HWM rayon–polyester blends with a durable-press finish show an abrasion resistance equal to similar blends containing cotton, although both are reduced. Some evidence has appeared which indicates that a polynosic rayon with a low level of resin can give a wash-and-wear rating equivalent to an HWM rayon at a normal durable-press level and the polynosic retains more resistance to abrasion. The method and the test conditions for determining abrasion resistance are the factors that make interpretation of results and comparison of data from other laboratories difficult. Nuessle and Gagliardi(*34*), in studying resin-treated and untreated rayons, found that as the severity of the test was reduced so was the difference in abrasion resistance between the two fabrics. Under very mild test conditions, the abrasion resistance of the resin-treated fabric was better than that of the untreated fabric. It was concluded that in their test method fiber elasticity was important.

Although limited work has been published on the location of the crosslinks in a rayon, some evidence exists which indicates that locations can vary. Park(*35*), using tritium-labelled urea–formaldehyde, found uniform distribution with no surface buildup, but also found a distribution difference between skin and core when using a tire yarn with a very thick skin. In a similar study, Ford(*36*) found that autoradiographs were not sufficiently clear to differentiate between skin and core crosslink locations. Using a staining technique it was shown that the core had taken up the resin preferentially. Nonuniform distribution has also been observed when crimped fibers are resin treated. The crimped fibers have surface areas that show no skin or only a very thin skin and these areas absorb more resin.

The core crosslinking technique of Willard, Tesoro, and Valko(*37*) on cotton has also been applied to rayon. This technique consists of treating the fabric with an aqueous solution of DMEU and zinc nitrate, drying and extracting the fabric with ethanol for varying periods of time, and finally curing. A cotton fabric will show a slightly lower crease recovery angle than an unextracted fabric after curing, but its abrasion resistance is improved considerably. On the other hand, rayon fabrics treated in this manner did not show a higher level of abrasion resistance. Lauchenauer and coworkers(*38*), however, using ammonia gas to deactivate the surface catalyst, found improved abrasion resistance for a polynosic fabric.

The technique of crosslinking rayon fibers in a wet or swollen state (Form W) has shown some interesting results when a cold solution is used(*39*). A regular rayon, given a Form-W treatment at low temperatures, will have a very high wet resilience, about normal dry resilience, a high fluid-holding capacity, and a high resistance to NaOH. Its strength under standard conditions will be 10–30% lower and its elongation will be 150–190% higher than an untreated rayon. If these fibers are then dry crosslinked using an acid catalyst cure, they develop a high dry resilience, are not brittle, have a lower fluid-holding capacity, and high caustic resistance. The tenacity falls between that of the original fiber and the wet crosslinked form, and the elongation is lower than the original fiber. As the linear density of a regular rayon is increased ($1\frac{1}{2}$ den. $\rightarrow$ 8 den.), less improvement in resiliency is obtained with some increase in brittleness. Subjecting a high wet-modulus fiber to this treatment produces higher resiliency than a regular rayon, and no brittleness is obtained.

B. FLAMEPROOFING

The conventional flameproofing agents discussed in the chapter on cotton are applicable for after-treating rayon fabrics also. With the increasing need for this property in fibers due to Federal regulations, attention has been given to the development of a flame-retardant rayon by incorporation of the flame-retardant compound in the fiber during its production. The phosphorus-containing resins which have been applied to cotton fabrics and which impart good flame resistance and durability in laundering are also applicable to rayon fibers and fabrics. However, whether flame retardancy is achieved by a topical application of chemicals or by incorporation of the flame-retardant chemical throughout the fiber cross section, such products have found limited acceptance at the consumer level. This might be due to the loss of certain aesthetic characteristics of the treated fabrics as well as the reduction in tensile strength, tear strength, and abrasion resistance. Lastly, the problem of the relatively high cost of such a fabric and lack of consumer recognition of the protection advantage of flame-retardant fabrics offer the textile industry an additional challenge for product development.

Within the past few years, the work on the reaction of tetrakis (hydroxymethyl) phosphonium chloride with reactive nitrogen compounds and telomers from bromoform and triallyl phosphate, brominated allyl esters of phosphonitrillic chlorides, and more recently, the tetrakis(hydroxymethyl) phosphonium hydroxide, have produced good flame-retardant compounds. Progress in the area of flame retardancy by the incorporation of the flame-retardant chemical in a fiber such as rayon has been slower. This may be explained by considering the requirements of such a chemical for effective survival in the rayon spinning process (*40*).

(1) The compound cannot be water soluble.
(2) A compound soluble in the alkaline cellulose xanthate solution could be used, provided it is insolubilized in acid solution (spin bath). However, the compound should not be soluble in weak alkali or it would be removed in dilute alkaline washing steps.
(3) Water dispersible and polymeric compounds could be incorporated if the molecular weight is higher than 50,000 since lower molecular weights can be leached from the fiber.
(4) Water- and alkali-insoluble compounds which can be dispersed as a pigment could be considered. However, over 5% solids in the viscose (alkaline cellulose xanthate solution) could result in spinning difficulties and a deterioration of fiber properties.
(5) The flame-retardant compound should be chemically stable in the alkaline and acid conditions of rayon spinning.
(6) The flame-retardant compound should have a relatively low vapor pressure or be nonvolatile at ambient or fabric finishing temperatures.
(7) Probably the compound should not be reactive with other chemicals encountered in fabric finishing processes.

One flame-retardant rayon is commercially available (*41*). Its development was on the basis of the above considerations. This fiber in staple form is stated to maintain its flame retardancy through peroxide or hypochlorite bleaching, dyeing, and resin treatment. The flame retardance is not lost by treatment with alkali, high-temperature laundering, or dry cleaning.

Godfrey (*42*), in his study of flame-retardant rayons, used differential thermal analysis (DTA) and thermogravimetric analysis (TGA) to elucidate pyrolytic behavior. The rayons were prepared by adding the compounds to the viscose before spinning the fiber. Included in

TABLE II
DTA OF FIBERS IN AIR

Compound	Conc., %	Exotherm begins, °C	Peaks °C
Control	–	300	339,469
Tetrakis(hydroxymethylphosphonium) chloride–thiourea polymer	14	260	295,300
Tris(2,3-dibromopropyl) phosphate	8	250	300,472
Mixed isopropoxyphosphaxenes	7	250	300,495
Hexapropoxycyclotriphosphazatriene	10	250	266,268, 300,500
Mixed propoxyphosphazenes	9.5	252	270,275, 300,500

this study were (see Table II): (1) an alkali-soluble polymer of tetrakis (hydroxymethyl) phosphonium chloride and thiourea; (2) tris(2,3-dibromopropyl) phosphate; (3) 2,2,4,4,6,6-hexaisopropoxycyclotri-phosphazatriene; (4) 2,2,4,4,6,6-hexapropoxycyclotriphosphazatriene; and (5) a product comprising mixed linear and cyclic propoxyphos-phazenes.

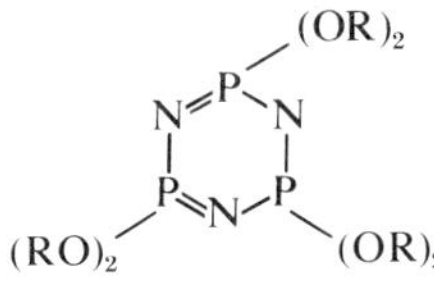

R = propyl, isopropyl

A cyclotriphosphazatriene

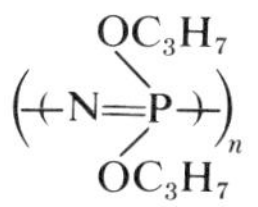

Linear propoxyphosphaze ne

The rayons were prepared so that although the fibers had moderate flame retardancy the ratio of cellulose to flame retardant was high. In this way, the influence of the flame retardant could be observed more clearly since the fibers remained basically cellulose. Thermograms show at what temperatures the main weight losses occurred (see Table III). Also as in all flame-retardant cellulosics, the residue remaining at 500°C is much larger than the untreated control rayon. In support of the chemical theory of flame retardancy, explanations of the decom-position reactions may be proposed. The THPC–thiourea polymer

TABLE III

Compound	Temp., °C
Control	240–355
Tetrakis(hydroxymethylphosphonium) chloride–thiourea polymer	220–320
Tris(2,3-dibromopropyl) phosphate	220–290
Mixed isopropoxyphosphazenes	243–270
Hexapropoxycyclotriphosphazatriene	230–310
Mixed propoxyphosphazenes	260–310

under oxidizing conditions can, during its breakdown, form phosphorus oxides. The tris(2,3-dibromopropyl) phosphate liberates bromine radicals and the final product of pyrolysis is probably phosphoric acid. The alkoxyphosphazenes which are unusually stable, probably follow the thermal rearrangement described by Shaw and coworkers(43), that is, compounds are formed which are readily hydrolyzed and these hydrolysis products react with the cellulose, leading to the lower decomposition temperature observed for the fiber.

In conclusion, the quantity of flame retardant required in rayon varies between 15 and 20% on the weight of the fiber. This is dependent not only on the flame-retardant compound and its chemistry, but also on the linear density of the fiber and the construction of the fabric. For example, a 1.5-den. fiber would require more of the flame-retardant compound incorporated in it than a 5-den. fiber.

C. Miscellaneous

1. *Grafting*

As stated previously, all vinyl monomers which can be grafted onto cotton can also be reacted with rayon. However, very little data has been published and one would conclude that the grafting technology in its present state yields little product improvement for rayon fibers. Kuzmak(44) reports that an ethyl acrylate graft, while improving the crush recovery of rayon, introduces a wet soiling problem due to the inherent properties of the polymeric graft. Grafting of compounds such as acrylonitrile or styrene–acrylonitrile onto fibers can give fabrics of high wrinkle recovery, low water imbibitions, a wool-like hand, and generally good wash-and-wear properties. Unfortunately, acrylonitrile grafts on rayon can render the fiber or fabric sensitive to temperatures such as might be experienced in later finishing treatments. Elastomeric properties can be imparted to rayon fibers by grafting several hundred percent by weight of methyl, ethyl, or butyl acrylate (45). Best results were obtained by crosslinking the fibers with formaldehyde, swelling with zinc chloride, and grafting in an aqueous emulsion system by the ceric ion method. At low levels of grafting (30%) butyl acrylate improves the abrasion resistance of rayon considerably.

A variation in grafting techniques involves thiocarbonated alkali cellulose. The utility of the system has been demonstrated with xanthated wood pulp and "unregenerated rayon" fibers. When grafting is carried out on undried cellulose xanthate fiber using hydrogen peroxide and monomer, as on an ethylene sulfide derivative of cellulose and a ferrous or ferric salt with the hydrogen peroxide, a very rapid reaction occurs with a high add-on (46). With cellulose xanthate fibers, the high alkalinity of the fiber should be reduced by thorough washing in order to prevent inhibition of the reaction and to remove certain by-product impurities which appear to induce the formation of considerably homopolymers. The use of hydrogen peroxide has been found to be an initiator for grafting mixtures of styrene–methyl methacrylate to rayon fibers when the fibers were produced from viscoses containing additives such as dimethylamine (47). Fibers produced from a non-additive viscose gave little or no add-on and cotton cannot be grafted by this system.

Vapor phase grafting on rayon fibers produces a rapid rate of graft polymerization and less homopolymer formation than is usually experienced in solution grafting. For grafting acrylonitrile by this method, potassium persulfate has been found to be a good catalyst that is stable in solution and produces a fiber which is not off color (48).

2. Acetylation

The acetylation of rayon can be carried out in acetic anhydride vapor at an elevated temperature (140–150°C) usually with sodium or potassium acetate as a catalyst. This acetylated fiber (DS 1.5–2) has properties that are superior to the fiber produced from acetylated cellulose pulp which is dissolved in acetone–water and dry spun (49). The Japanese have marketed a product made by vapor phase acetylation of an all-skin rayon. The fiber has good strength (over 2.5 g/den.), is heat settable, has better wash-and-wear performance than cellulose acetate, and has about 5% moisture regain.

V.   Rayon Tire Cords

The introduction of rayon into automobile tires became important when rayons with sufficient strength to compete with cotton were first produced. This was accomplished by the introduction of zinc sulfate in the spin bath, of additives such as amines and ethoxylated phenols in the viscose and spin bath, and various stretching procedures in hot dilute acid baths. These changes, along with a better α-cellulose pulp and further understanding of the viscose process, have led to the fiber used in tires which is best described as a high-strength, medium elongation rayon. It has the usual rayon crystallinity but with smaller crystallites which are more highly oriented than in other rayons. They have a significantly lower water imbibition and lower cross sectional swelling but higher moisture regain values than normal rayon. Their chemical resistance is similar to other rayons except that they have a greater sensitivity to the action of alkali than the new high wet-modulus fibers. Specific details of the tire-yarn process are more fully discussed by Schappel and Bockno(4).

REFERENCES

1. British Patent 1,130,484, Rayonier, Inc.
2. Dunio Hata and Kingo Yokota, *J. Soc. Fiber Sci. Technol. Japan,* **24**, 420 (1968).
3. L. H. Phifer and John Dyer, *Macromolecules,* **2**, 118 (1969).
4. J. W. Schappel and G. C. Bockno, "Cellulose and Cellulose Derivatives" Part IV, Chpt. 8B, N. Bikales and L. Segal, eds., John Wiley and Sons, to be published.
5. G. C. Daul and T. E. Muller, *J. Appl. Poly. Sci.,* **12**, 487 (1968).
6. H. Mark, *Chemiefasern,* **15**(G), 422 (1965).
7. S. Poznanski, *Przem. Che.,* **22**, 463 (1938); U.S. Patent 2,312,152 (1943), General Aniline Film Corp.
8. John Dyer and L. H. Phifer, private communication.
9. D. K. Smith, *Text. Res. J.,* **29**, 32 (1959).
10. H. Klare and A. Gröbe, *Osterr. Chemiker-Ztg.,* **65** (7), 218 (1964).
11. I. H. Welch, *Am. Dyest. Rep.,* **50**, P25 (1961).
12. U.S. Patent 3,241,553, Johnson and Johnson.
13. I. Wizon, *J. Appl. Poly. Sci.,* Appl. Poly. Symp. No. 9, 395 (1969).
14. G. Casperson, H. Haeusel, and G. Hoffmann, *Faserforsch. Textiltech.,* **18**, 455 (1967).
15. Ronald E. Dehl, *J. Chem. Phys.,* **48**, 831 (1968).
16. F. P. Morehead and W. A. Sisson, *Text. Res. J.,* **15**, 443 (1945); D. N. Tyler and N. S. Wooding, *J. Soc. Dyers Colour.,* **74**, 283 (1958).
17. W. A. Sisson, *Text. Res. J.,* **30**, 153 (1960).
18. W. M. Kaeppner, *Text. Res. J.,* **38**, 662 (1968).
19. W. Kling, H. Mahl, and W. Heumann, *Melliand Textilber.,* **44**, 335 (1963).
20. A. Gröbe and H. J. Gensrich, *Faserforsch. Textiltech.,* **20** (3), 118 (1969).

21. R. H. Braunlich, *Am. Dyest. Rep.,* **54**, P160 (1965).

22. Technical Service Bulletin S-24, FMC Corp., American Viscose Div.

23. S. Pilichowska-Gwozdz and W. Lewaszkiewicz, *Przegl. Wlok.,* **22** (8), 403 (1968).

24. T. F. Cooke, J. H. Dusenbury, R. H. Kienle, and E. E. Lineken, *Text. Res. J.,* **24**, 1015 (1954).

25. R. A. Gill and R. Steele, *Text. Res. J.,* **32**, 338 (1962).

26. D. H. Morton and C. Beaumont, *J. Soc. Dyers Colour.,* **76**, 578 (1960).

27. Tibor Robinson, *Lenzinger Berichte,* **23**, 45 (1967).

28. K. H. Tauss, *Am. Dyest. Rep.,* **53**, 171 (1964).

29. U.S. Patent 2,960,484; 3,113,826, Courtaulds, Ltd.

30. L. Rebenfeld, *Text. Res. J.,* **34**, 168 (1964).

31. Technical Service Bulletin, S-40, FMC Corp., American Viscose Div.

32. Technical Service Bulletin, S-43, FMC Corp., American Viscose Div.

33. British Patent 1,144,604, J. P. Stevens and Co.

34. A. C. Nuessle and D. D. Gagliardi, *Amer. Dyest. Rep.,* **40**, 409 (1951).

35. G. S. Park, *J. Soc. Dyers Colour.,* **76**, 624 (1960); *ibid.,* **78**, 451 (1962).

36. J. E. Ford, *J. Soc. Dyers Colour.,* **77**, 209 (1961).

37. J. J. Willard, G. C. Tesoro, and E. I. Valko, *Text. Res. J.,* **39**, 413 (1969).

38. A. E. Lauchenauer, H. H. Bauer, P. F. Matzner, G. W. Toma, and J. B. Zuercher, *Text. Res. J.,* **39**, 585 (1969).

39. F. R. Smith, FMC Corp., private communication.

40. J. W. Schappel, *Mod. Text.,* **49** (7), 54 (1968).

41. U.S. Patent 3,455,713, FMC Corp.

42. Leonard E. A. Godfrey, *Text. Res. J.,* **40**, 116 (1970).

43. B. W. Fitzsimmons, C. Hewlett, and R. A. Shaw, *J. Chem. Soc.,* 1735, 4459 (1964).

44. J. M. Kuzmak, FMC Corp., private communication.

45. Yoshio Nakamura, Michiharu Negishi, and Toshiko Kakinuma, *J. Poly. Sci.,* Part C-2, No. 23, 629 (1968).

46. Belgian Patent 646,284; 646,285, Scott Paper Co; U.S. Patent 3,359,224, Scott Paper Co.

47. U.S. Patent 3,475,357, FMC Corp.

48. P. A. Smith, FMC Corp., private communication.

49. Takeo Takagi and J. B. Goldberg, *Mod. Text.,* **41** (4), 49 (1960).

Chapter 3 CELLULOSE ACETATE

I. Introduction

Cellulose acetate is a well-known derivative of cellulose and has found many uses as a fiber, film, and plastic coating. The Federal Trade Commission defines cellulose acetate fibers in the following manner: "*acetate* — a manufactured fiber in which the fiber-forming substance is cellulose acetate. Where not less than 92% of the hydroxyl groups are acetylated, the term *triacetate* may be used as a generic description of the fiber."

Cellulose acetate was not adopted as a fiber-forming polymer immediately upon its preparation because it was not readily dissolved in low-boiling organic solvents. In the early 1900's Miles and von Bayer discovered that solubility could be more easily achieved if some of the acetyl groups in a fully substituted cellulose acetate were removed by acid hydrolysis. This discovery was commercialized by Camille and Henri Dreyfus. Secondary cellulose acetate fibers have been commercially important since World War I; however, cellulose triacetate did not appear as a fiber until the late 1940's and only became commercial in 1954. Acetate and triacetate fibers are similar in chemical structure with acetate having about 83% of the hydroxyls acetylated. There are some important differences in fiber properties, particularly those connected with the response of the polymer to elevated temperatures.

II.   Preparation of Cellulose Acetate

The starting material, cellulose, may be either cotton linters or a specially purified wood pulp with a high α-cellulose content. Cotton linters were used originally, but their cost and availability led the industry to adopt wood pulp as the principal cellulose source for acetylation. Careful quality control of the purity, molecular weight, and molecular weight distribution of the cellulose is maintained by the pulp manufacturer.

The acetylation reaction is quite simple chemically and may be visualized to proceed in the following manner:

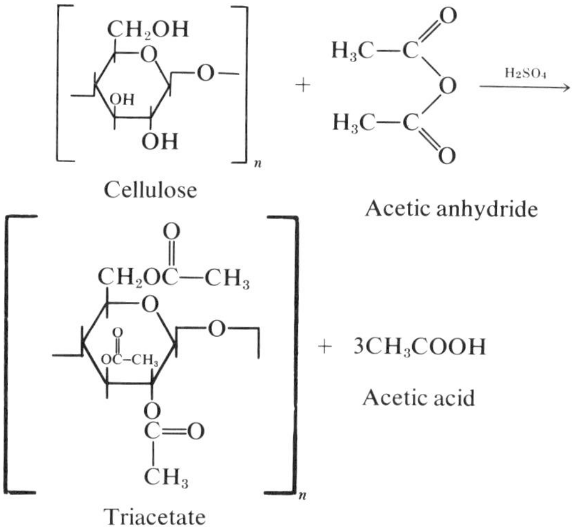

Cellulose

Acetic anhydride

Triacetate

The above equation shows that the three hydroxyls on the repeat glucopyranose unit have reacted. This can occur only if the hydroxyls are accessible to the reactants. In practice the wood pulp is shredded and preswollen with acetic acid to achieve a uniform reaction. At the beginning of acetylation, substitution occurs in a somewhat random manner because of the limited accessibility of the hydroxyl groups. At this time in the reacting matrix there exist cellulose chains whose glucopyranose units are substituted partially, as shown below. The three hydroxyls may be ranked in order of decreasing reactivity in the acid catalyzed acetylation reaction as C-6, C-2, and C-3. Therefore, if the polymer molecules were completely accessible, the principal partially acetylated units would be **I**, **II**, and **III**, although some quantity of every partially acetylated species exists.

The usual procedure for obtaining secondary acetate commercially is to esterify the hydroxyls completely (cellulose triacetate) and then to hydrolyze back partially to the desired degree of substitution. The hydrolysis is usually carried back to a degree of substitution of about •2.5. Commercially, the acetylation reaction always begins in a heterogeneous medium after the acetic acid preswelling treatment. Other preswelling chemicals, such as glycol ethers, have been evaluated, but acetic acid is preferred for performance and cost. The catalysis of the reaction may be achieved by acids other than sulfuric and include sulfamic and hydroxylamine sulfate(*1*). Malm, Tanghe, and Schmitt(*2*) have studied reaction rates using, also, perchloric, methanedisulfonic, sulfonacetic, and methansulfonic acids. The purpose of their work was to determine why sulfuric acid was a preferred catalyst for the acetylation of cellulose, and their conclusions were (1) that it concentrates at the hydroxyls, (2) that it reacts with cellulose forming —OSO_3H groups, and (3) that this cellulose acetate sulfate triester is more soluble in the acetylating bath than cellulose triacetate. Thus, as acetylation proceeds, the reaction mass changes from shredded fibers through an opaque, stringy or doughy stage to a very viscous, clear solution. The "fully" acetylated cellulose, having slightly less than the theoretical acetyl content of 44.8%, is then subjected to controlled hydrolysis treatments to remove the sulfate groups as well as some of the acetate groups. This is accomplished by the addition of water, which is usually added as dilute aqueous acetic acid. In this manner, local precipitation of the product is prevented. The rate of removal of these groups is controlled by time, temperature, and acidity of the solution. The sulfates are removed first under relatively strong acid conditions and some hydrolysis of the polymer occurs, thus lowering the degree of polymerization. The removal of the acetate groups occurs more slowly; usually, the temperature is increased to facilitate this reaction. To stop this hydrolysis at the desired acetyl content, sodium acetate, or perhaps magnesium acetate, is added to neutralize the sulfuric acid.

The kinetics of the acetylation of cellulose (*3*) and the acid-catalyzed degradation of cellulose acetate (*4*) have been studied, and the importance of diffusion, concentration of reactants and catalyst, and temperature are shown. Both the esterification and the hydrolysis to the desired acetyl content are accompanied by considerable degradation of the cellulose. This degradation must be carefully controlled in order for the polymer to be useful in a fiber form.

At the end of the reaction, the cellulose acetate is recovered by precipitation from solution with water. This precipitate, called acetate flake, is washed thoroughly until neutral and dried to the desired moisture content.

The commercial process for preparation of cellulose triacetate is similar to that for cellulose acetate except for the partial hydrolysis at the end of complete acetylation. Continuous acetylation processes are possible for the production of the triacetate and secondary acetate, and are desirable if the spinning of fibers occurs at the same plant site.

III. Cellulose Acetate Fibers

A. SPINNING

Secondary acetate fibers are prepared by a dry-spinning method using acetone containing approximately 4% water as the solvent. The water is added to the acetone to decrease solution viscosity which passes through a minimum at a 10–13.5% water content. The cellulose acetate concentration in this solvent ranges from 20–25%. The properties of this viscous solution vary considerably with molecular weight, degree of esterification, and the chemical and physical uniformity of the cellulose acetate (*5*).

The cellulose acetate solution is filtered and deaerated before spinning. In the dry spinning process the viscous solution is extruded through a spinneret downward into a heated vertical column. The flow of polymer solution downward is countercurrent to the passage of hot air, used to evaporate and carry away the solvent. The take-up rate onto the filament package as compared to the pump speed to the spinneret determines the linear density and the extent of drawing. Fiber structure and properties are largely determined at the incipient stage of formation in the heating zone and precise control of speeds, temperatures, and immediate environment are essential.

Cellulose acetate fibers may be delustered by the addition of titanium dioxide to the spinning dope, and other additives such as fire retardants may be included also. As the fibers emerge from the spinning tube a finish is applied for lubrication to aid further processing.

B. FIBER PROPERTIES

1. *Physical*

The cellulose acetate fiber, commonly known as acetate, is characterized by a relatively low degree of orientation and crystallinity. The fiber properties are given below.

Tenacity, g/den.	1.2–1.5
Elongation, %	25–45
Recovery (at 4% elongation)	45–65
Moisture regain, (std. conditions)	6–6.5
Specific gravity	1.32

The strength decreases to about 0.65–0.75 g/den. and elongation increases slightly when the fiber is wet. Tensile properties of acetate fibers can be changed by stretching and drawing operations after spinning. In general these will tend to increase fiber strength and stiffness, and decrease its elongation. This fiber, in common with other textile fibers, is viscoelastic, displaying creep and a strong time dependency in its stress–strain behavior. It has a low resistance to abrasion and shows thermoplasticity at 300°F with softening between 380 and 400°F.

2. *Chemical*

Acetate fibers are resistant to peroxide bleaches and to chlorine bleaches under either slightly acidic or basic conditions. They are attacked by strong oxidizing agents and hydrolyzed by strong acids. In the presence of alkali the acetate groups are removed by saponification leaving, essentially, a regenerated cellulose fiber. The acetate fiber is soluble in acetone, and swells or partially dissolves in acetic acid and other organic solvents. However, the dry-cleaning solutions normally used do not affect the fiber.

Acetate fibers generally are more stable to sunlight than nylon, rayon, and cotton but less stable than acrylic and polyester fibers. Egerton and Shah(6) studied the combined effects of moisture and temperature in the presence and absence of TiO_2, and concluded that TiO_2 accelerated photochemical degradation in the presence of moisture, particularly at elevated temperatures. Acetate also showed considerable strength loss in moist air at temperatures above 7°C, even when free of TiO_2.

C. MODIFICATIONS

1. *Additives*

Solution or dope dyeing is common with acetate fibers. Organic or inorganic pigments are added to the spinning dope and the cellulose acetate is extruded as colored filaments. By incorporation of the proper chemicals in the spinning dope, acetate fibers may be made fire retardant, more stable to uv degradation, less subject to delustering by heat, and more resistant to the building up of static. For fire retardancy, compounds, such as bis(bromochloropropyl) bromochloropropyl-phosphonate(7) and amphoteric metal pyrophosphates(8), may be dissolved with the cellulose acetate in acetone before spinning. Alternatively, the fire-retardant compound may be incorporated in the size applied to the fiber. One efficient size is a copolymer of styrene–maleic anhydride derivatized with ethylene bromohydrin(9). For protection against uv radiation phenyl iodosalicyclic acid esters(10) are effective, as are a variety of aromatic heterocyclic compounds(11). Acetate fibers reacted with isocyanates to form carbamates on the remaining free hydroxyl groups also show improved resistance toward uv radiation(12). The loss of luster by heat is probably more of a problem in acetate fibers than in triacetate fibers. This can be minimized in both fibers by the incorporation of a dicarboxylic acid such as the phthalic acids or adipic acid(*13*). Prolonged treatment of the cellulose acetate flake with dilute boiling solutions of sulfuric acid(*14*) will also yield a fiber resistant to thermal delustering. The reduction of static is most often accomplished by a surface treatment of the fiber, although antistatic agents can be added to the spinning dope. Among the many useful compounds are the quaternary ammonium salts of long-chain aliphatic alcohols(*15*). Aftertreatment compounds include ethylene oxide adducts of phenolic compounds(*16*), 2-carboxybutyl-propylphosphonic acid(*17*) derivatives of ethylenediamine(*18*), or a reactive polymer such as hydroxyethyl cellulose and the disodium tris(β-sulfatoethyl)-sulfonium salt(*19*). For certain end uses, the acetate fibers may require plasticization which may be accomplished by incorporation of glycerol triacetate, polyethylene glycol diacetate (*20*), or dialkyl sulfones(*21*).

2. *Grafting*

Cellulose acetate with a degree of substitution of 2.3–2.5 has a sufficient number of unreacted hydroxyl groups to allow it to undergo the typical grafting reactions of cellulose. Monomers which have been used in grafting include styrene, acrylonitrile, many acrylate and methacrylate esters, and vinyl acetate. Grafts may be obtained by free-radical initiation from redox and peroxide systems or from high-energy radiation, such as γ rays. Among the acrylate grafts obtained by a ceric ion initiator and vapor phase treatment with the monomers, improved dry- and wet-crease recovery was obtained with propyl and butyl acrylate(*22*). About 5% graft gave maximum improvement in crease resistance and tear strength. The fibers also showed rubber-like elasticity and water repellency. Polystyrene grafted fibers have been studied extensively but show no desirable property improvements(*23*). Hamburger(*24*) used styrene and cellulose acetate in studying the effects of retarding agents, such as nitrobenzene, on the chain growth. He found that the degree of polymerization (DP) of the polystyrene side chain was always higher than that of the homopolymer in solution, indicating that termination is diffusion controlled within the swollen substrate. When the acetate was slightly swollen, relatively large amounts of nitrobenzene were required for significant reduction of DP. If the substrate was highly swollen, then only trace amounts were required to reduce DP. Rogovin(*25*) also reviewed the effect of a regulator in grafting 2-methyl-5-vinylpyridine on acetate, and a significant reduction of molecular weight was obtained. However, from a commercial point of view, grafting has not yet provided sufficient improvement in fiber properties to make this modification technique attractive.

IV. Cellulose Triacetate

A. SPINNING

Cellulose triacetate is converted to fiber commercially by a dry-spinning method similar to acetate spinning, although a different solvent is used. The triacetate polymer dissolves readily in mixtures of methylene chloride and methanol, and such a solution is used commercially. Another solvent mixture suitable for spinning is methyl acetate–acetone(26). In preparing the spinning solution, a delusterant such as titanium dioxide is added if such a property is desired in the final fiber. The wet-spinning method(27), which produces a fiber about twice as strong as the dry-spinning method, is no longer used commercially.

A lubricant is applied after fiber formation and is usually based on a white mineral oil. Lubricant formulations may also include fatty acid glycerides, polypropylene or polyethylene glycols, or silicones.

B. FIBER PROPERTIES

1. *Physical*

The principal differences in fiber properties between secondary acetate and triacetate are those associated with the fact that cellulose triacetate can be heat set. In view of the structural regularity of the fully acetylated cellulose chains, the polymer can be made to crystallize under the combined effects of heat and mechanical constraint. Heat setting can be accomplished at 240°C in 30 sec, at somewhat lower temperatures in longer times, or with steam at about 125°C. The development of crystallinity by heat setting of cellulose triacetate can be clearly seen in x-ray diffraction patterns(28). Typical fiber properties are listed below:

Tenacity, g/den.	1.2–1.4
Elongation, %	25–35 filament
	34–40 staple
Moisture regain, %	2.5–3.0
Initial modulus, g/den.	44

Specific gravity	1.32
Melting point	570°F
Tensile recovery, %	50–65 at 5% stretch

Triacetate shows many of the characteristics of the synthetic fibers in that it has quick drying, good easy-care properties and may be permanently pleated in the heat-setting treatment. The uniformly substituted cellulosic polymer has a typical glass transition temperature (186°) and crystallization temperature (213°C)(29). Its melting point is higher than acetate, occurring at approximately 309°C. These thermal properties permit a higher safe-ironing temperature and provide greater resistance to aqueous systems at elevated temperatures than acetate. After dry-heat setting, an atmospheric steaming may be given the fabric to release strains. Heat setting produces a higher softening temperature, a reduced water absorbency, and improved dimensional stability(30, 31).

2. Chemical

The resistance of cellulose triacetate to most solvents is good, although it softens in trichloroethylene. It is degraded or destroyed by such chemicals as hydrochloric acid, ammonium and sodium hydroxide, phenol, and sodium hypochlorite (5%). Its resistance to degradation by chlorine bleaches as normally used in the home is high. Degradation by uv light is comparable to acetate. After uv radiation for 20 h (no glass filter) the viscosity decreased by 65% and the free acid increased by 8.25%(32). However, the quantity of free acetic acid extracted depended both on the temperature of the water and the time of extraction. No suggested mechanism was put forward for these observed changes. A possible pathway might be the formation of a free radical with termination being brought about by a molecule of water, thus regenerating an —OH on the glucopyranose ring and producing acetic acid.

C. MODIFICATIONS

The finishing of triacetate fibers is not particularly broad in scope and mainly includes heat-setting techniques, treatments to improve the elastomeric properties, and surface saponification to reduce static electricity buildup. Grafting or crosslinking reactions have not been explored due to the small number of reactive hydroxyls available in the fiber. However, Kubushiro, Takemoto, and Imoto(33), among others, report that the triacetate does initiate the polymerization of methyl methacrylate in the presence and absence of water, but no explanation for this has been offered. It is better to dye the fabric before heat treatment thus obtaining not only increased dimensional stability but also better fastness of the dyes(*34*). The chemical and physicochemical changes in triacetate by dry heat have been carefully studied(*35*) and only structural changes occur in the fiber up to approximately 200–210°C, but above these temperatures (220°C), chain scission predominates as well as an increase in carbonyl groups. Ironing temperatures may be raised by a heat treatment in aqueous benzyl alcohol(*36*), or with carboxylic esters, ethers, or ether carboxylic esters, such as diethylene glycol diacetate(*37*).

Elastomeric fibers can be obtained by treatment with β-butoxyethyl acetate, acetyl triethyl citrate, and benzyl alcohol(*38*). A low molecular weight cellulose acetate may be linked to polyesters or polyethers with a diisocyanate such as 4,4′-methylenebis(phenyl isocyanate) to form a urethane-type polymer which, when spun, shows good stretch properties(*39*).

Surface saponification of these or secondary acetate fibers yields a thin film of regenerated cellulose which imparts antistatic properties as well as producing reactive sites for durable finishes and improving the ease of soil removal(*40*). Most likely this mild saponification begins with the removal of the primary hydroxyl groups(*41*) on the fiber surface. Both acetate and triacetate fibers have been reacted with tolylene diisocyanate by a vapor phase method(*42*). For both fibers the breaking strength remains almost constant and the elongation decreases. The initial modulus and elastic recovery increases.

REFERENCES

1. A. Takahashi and S. Takahashi, *J. Soc. Fiber Sci. and Tech. Japan.* Polymer Report No. 136, 29 (1969).

2. C. J. Malm, L. J. Tanghe, and J. T. Schmitt, *Ind. Eng. Chem.,* **53**, 363 (1961).

3. C. M. Conrad, P. Harbrink, and A. L. Murphy, *Text. Res. J.,* **33**, 784 (1963).

4. A. J. Rosenthal, *J. Poly. Sci.,* **51**, 111 (1961).

5. F. L. Wells, W. C. Schnattner, and A. Walker, *TAPPI* **46**, 581 (1963).

6. G. S. Egerton and K. M. Shah, *Nature,* **202**, 81 (1964).

7. U.S. Patent 3,321,330, FMC Corp.

8. U.S. Patent 2,989,406, Eastman Kodak Co.

9. U.S. Patent 3,454,588, FMC Corp.

10. French Patent 1,455,699 Societe Rhodiaceta.

11. Swiss Patent 410,851, J. R. Geigy A.G.

12. U.K. Belyakov, A. A. Berlin, V. P. Dubyaga, L. V. Nevskii, and O. G. Tarakanov, *Plast. Massy,* **1968** (8) 35.

13. U.S. Patent 3,369,916, Rhone-Poulenc S.A.

14. German Patent 1,811,531, Teijin Ltd.

15. U.S.S.R. Patent 186,619, All-Union Scientific Research Institute of Synthetic Fiber.

16. U.S. Patent 3,333,983, Nopco Chemical Co.

17. U.S. Patent 3,300,337, Stauffer Chemical Co.

18. V. I. Luzan, V. A. Blinov, K. A. Kornev, and T. M. Aleksandrova, *Khim. Prom. Ukr.,* **1966** (4) 14.

19. British Patent 1,059,568, I.C.I. Ltd.

20. British Patent 1,158,979, Eastman Kodak Co.

21. U.S.S.R. Patent 238,147.

22. K. Suzuki, I. Kido, S. Fujii, and Y. Inoue, *Sen-i Gakkaishi.* **23**, 559 (1967).

23. K. Hayakawa, K. Kawase, and T. Matsuda, *Nagoya Kogyo Gijutsu Shikensho Hokoku.* **15** (1), 21 (1966).

24. C. J. Hamburger, *TAPPI.* **50**, 510 (1967); *J. Poly. Sci..* **7A**, 1023 (1969).

25. Z. A. Rogovin, *Pure & Appl. Chem..* **14**, 523 (1967).

26. British Patent 857,390, DuPont Co.

27. U.S. Patents, 2,999,004; 3,057,039; 3,071,807; 3,081,145; 3,084,414; 3,109,697; 3,133,136; Belgian Patent 612,915, Celanese Corp.

28. S. B. Sprague, J. L. Riley, and H. D. Noether, *Text. Res. J..* **28**, 275 (1957).

29. J. K. Gillham and R. F. Schwenker, *Appl. Poly. Symp.* No. 2, 59 (1966).

30. M. W. Alford, *J. Text. Inst.,* **52**, P242 (1961).

31. W. Fester and S. T. Liu, *Textil-Praxis.* **21**, 440 (1966).

32. K. Kubrishiro, K. Takemoto, and M. Imoto, *Bull. Chem. Soc. Japan.* **42**, 3327 (1969).

33. A. Paulauskas and R. Leparskyte, *Tekst. Prom. (Moscow),* **27**, (7), 19 (1967).

34. S. Okuneckis, *Rayonne Fibres Syn..* **22**, 361, 441 (1966).

35. W. Fester and S. T. Liu, *Tex.* **26** (3), 180 (1967).

36. British Patent 903,971, British Celanese, Ltd.

37. British Patent 856,729, British Celanese, Ltd.

38. U.S. Patent 3,447,806, Celanese Corp.

39. U.S. Patent 3,386,930, Celanese Corp.

40. F. Fortess and A. F. Tesi, *Textil-Rdsch..* **16**, 745 (1961).

41. R. G. Zhbankov, R. V. Zueva, L. V. Savel'eva, and P. V. Koglov, *High Molecular Weight Compounds,* **2**, 1270 (1960).

42. K. Suzuki, I. Kido, M. Hyo, and S. Inoue, *J. Soc. Fiber Sci. and Tech. Japan.* **25**, 278 (1969).

Chapter 4 **WOOL**

I. Introduction

Wool is an animal hair from the body of sheep. This hair is sheared annually and its quality and quantity varies widely depending on the breed of sheep and environment. It is a mammalian hair belonging to a family of proteins, keratins, whose chemical structure has not been fully characterized.

When removed from the sheep, wool contains wax, perspiration, dirt, and extraneous vegetable matter. These impurities must be removed before the wool can be used for textile purposes by a number of processes including carbonization, scouring, and bleaching. Sheep are broadly classified by the types of wool that they produce, although a classification such as fine, medium, long, and crossbred does not adequately describe the sheep and their wool. Wool fibers differ not only in gross characteristics, such as diameter and length, but also in mechanical properties, crimp, scaliness, and color. It is a fiber with considerable structural heterogeneity on which extensive studies have been carried out. Research is developing a rather sophisticated picture of fiber structure, but there is not complete agreement on the interpretation of available data.

II. Structure

A. CHEMICAL

The building blocks of a wool fiber are α amino acids whose general formula is $H_2N-C(H)(R)C(=O)OH$. The amino acids are arranged in a polypeptide chain in the following manner:

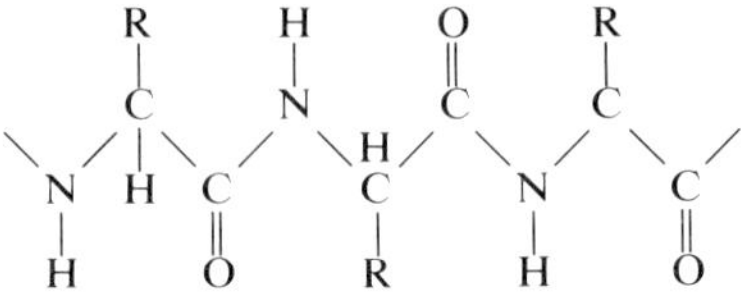

Table I lists the amino acids in wool according to the chemical character of the R group. The polypeptides containing these amino

TABLE I
AMINO ACIDS FOUND IN WOOL

Name	Structure of side chain (—R)
Hydrocarbon	
Glycine	—H
Alanine	—CH_3
Valine	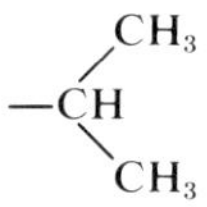
Leucine	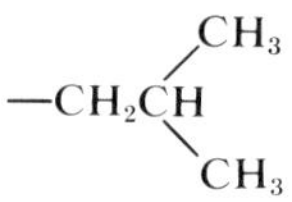
Isoleucine	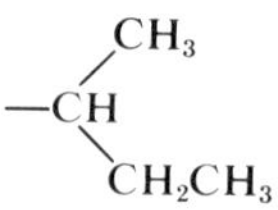
Phenylalanine	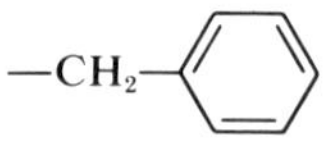
Proline (cyclic, "other" end connects to N atom)	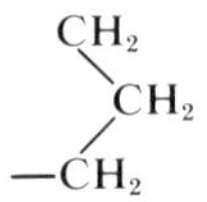

TABLE 1 (Cont.)

Name	Structure of side chain (—R)
Hydroxy	
Serine	—CH$_2$OH
Threonine	—CH with OH and CH$_3$
Tyrosine	—CH$_2$—〈benzene ring〉—OH
Acidic (free and as amide)	
Aspartic acid	—CH$_2$C(=O)—OH
Glutamic acid	—CH$_2$CH$_2$C(=O)—OH
Basic Arginine	—CH$_2$CH$_2$CH$_2$N(H)—C(=NH)NH$_2$
Lysine	—CH$_2$CH$_2$CH$_2$CH$_2$NH$_2$
Hydroxylysine	—CH$_2$CH$_2$CH(OH)CH$_2$NH$_2$
Histidine	—CH$_2$—C〈imidazole ring: NH—CH, CH—N〉
Heterocyclic Tryptophan	—CH$_2$—〈indole ring, N—H〉
Sulfur containing Cystine	—CH$_2$—S—S—CH$_2$—CH(⁺NH$_3$)(CO$_2^-$)
Methionine	—CH$_2$CH$_2$—S—CH$_3$

acids form a complex protein called keratin, differing somewhat in chemical structure. The low-sulfur fraction, known as S-carboxy-methylkeratin A (SCMKA), is found predominately in the organized portion of the fiber, the microfibril. The high-sulfur fraction, called S-carboxymethylkeratin B (SCMKB), is believed to be in the matrix, the structurally disorganized material surrounding the microfibrils. The polypeptide chains are flexible with free rotation around single bonds, and in the microfibrils they exist in a helical configuration.

The important intermolecular forces which stabilize the α-helix structure in the microfibrils are disulfide bonds, hydrogen bonds, ionic bonds (salt linkages), van der Waals forces, and hydrophobic bonds. Presumably, in the matrix the polypeptide chains exist in a random coil form where some of these intermolecular forces are also operative. It is believed that an important contribution of the disulfide bonds is in bridging the polypeptide chains of the microfibrils and the matrix. Some of the intermolecular forces are illustrated below.

(1) Covalent linkage

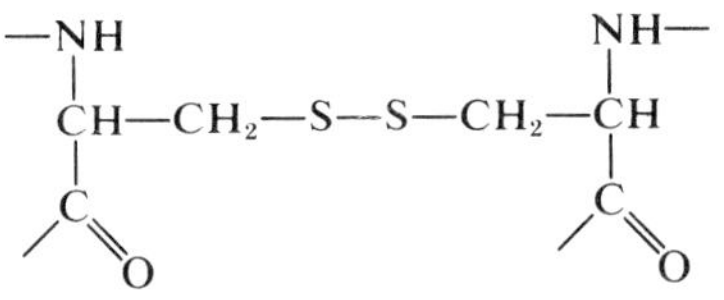

Disulfide crosslink

(2) Hydrogen bonds

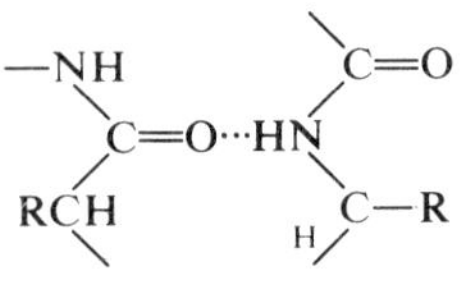

(3) "Salt" linkage

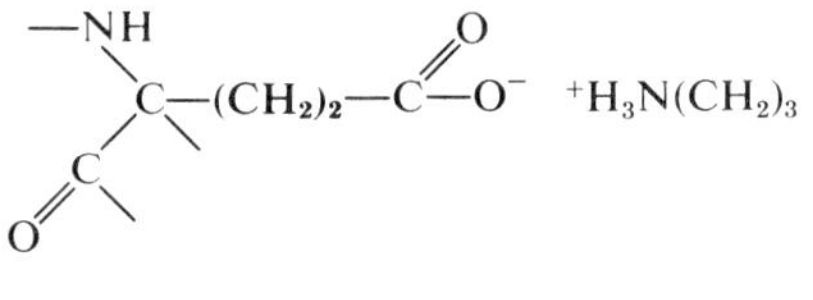

Glutamic acid moiety   Lysine moiety

B. PHYSICAL

The wool fiber is morphologically complex and its characterization has been aided by light and electron microscopy, x-ray diffraction, and many other techniques of structural analysis. Grossly, the fiber consists of: (1) cuticle, the somewhat irregular scales on the surface; (2) epicuticle, a thin membrane located on the surface of the scales; (3) cortex, inner portion encompassing about 90% of the fiber; (4) medulla, a central core arising from the growing root, more often observed in the coarser wools. The cuticle and epicuticle play an important role in many fabric properties and in controlling the rate of diffusion of dyes and other molecules into the fiber(1). The cortex, the predominant portion of the fiber, controls the bulk properties of the wool. The cortex of crimped wool fibers has a bilateral structure and is composed of two types of cells, referred to as ortho and para(2). These cells differ in chemical composition and density, and may be differentiated by selective staining techniques. The ortho-cortex cells are chemically more reactive and have the greater receptivity to certain dyes. Figure 1a illustrates this bilateral structure which is responsible for the crimp.

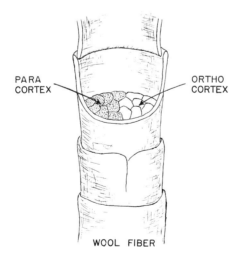

PARA CORTEX

ORTHO CORTEX

WOOL FIBER

FIG. 1a.   Pictorial drawing showing the bilateral structure of a wool fiber.

The cortical cells have a substructure containing microfibrils, which are approximately 80 Å in diameter and several microns long. These, in turn, are composed of protofibrils. Latest nomenclature seems to be to refer to the microfibrils as filaments containing protofilaments(*3*). The heterogeneous keratin in the cortical matrix consists of high-sulfur proteins(*4*) and seems to have a lower molecular weight than the low-sulphur proteins in the microfibril.

The protofibrils appear to be 20 Å in diameter(*5*) and from x-ray data (to account for the 1.5 and 5.1 Å meridional reflections), Crick(*6*) has proposed a coiled, coil-rope model suggesting a three stranded rope structure, although the possibility of a doublet is not denied(*7*). Pauling and Corey(*8*) in analysis of their x-ray data, have suggested the possible concept of a cable model. Johnson and Speakman(*9*) attempted to isolate protofibrils whose structure would be unchanged for study with high-resolution electron microscopy. They were unable to confirm by direct observation that protofibrils are triplets of the Crick coiled coils, but the dimensions of the predominant units which were obtained were consistent with the three-strand rope arrangement (Fig. 1b).

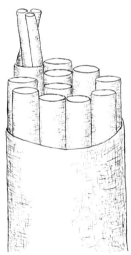

FIG. 1b.   Pictorial drawing showing the three-strand rope structure in a protofibril.

From the known lateral dimensions of the microfibril, the maximum number of protofibrils of which it can consist are 11. Bendit(*10*) has reviewed various models for the protofibrils and considers, along with Crewther and coworkers(*11*), a 9 + 2 arrangement in the micro-fibril, a feasible working model, with three α helices (rope structure) in each protofibril. However, structures are still under study and in spite of the extensive discussions in the literature, no completely definitive structural model is yet available.

III.   Physical Properties

Fiber fineness is an important property in determining the grade or count given to wool "lots." Normally, wool fibers range in diameter from 16 μ in the fine merinos to over 40 μ in the coarsest wools. Cross sectional shape can vary from an oval to a circle and the frequency of crimps varies also. Staple lengths, which generally increase with increasing fiber diameter, may range from $1\frac{1}{2}$ in. in fine wools to over 8 in. in coarse wools. The luster of the fiber, which is dependent on the structure of the surface and on the size and straightness of the fiber, is generally described as *silver*, *silk*, and *gloss*. A mild silver luster would describe a fine, highly crimped merino wool; a silk luster is found in long-stapled, long-waved wools; and a glossy luster occurs in straight, smooth fibers sometimes found on the head, neck, and legs.

The stress–strain properties of wool have been extensively studied and interpretations of the fiber's response under different conditions become complex due to its complicated chemical and physical structure. Figure 2 shows a schematic stress–strain curve for a single wool fiber at a constant rate of strain. In terms of the native configuration and structural arrangements, an interpretation of this deformation concerns principally the covalent disulfide bond

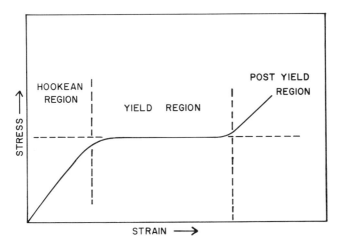

Fig. 2.   Schematic stress–strain curve of wool fiber.

and the transformation of the α-helix to an unfolded form called the β configuration. In the Hookean region, the stress is essentially linear to the strain reflecting bond stretching and bond-angle deformation. At about 1–2% strain, the fiber yields and the α-helix configuration begins to unfold(*12, 13*). From the flatness of the curve in the yield region, it is evident that this unfolding process requires only a very small amount of force. However, the unfolding does not go to completion, and at an extension of ~30%, the stress increases sharply. The fiber breaks at an extension of 50–60% which is before $\alpha \rightarrow \beta$ transformation is complete. The structural and molecular interpretation of the postyield region is under considerable controversial discussion (*14*). It has been shown that modification of the disulfide content or changes in the —SH content have a strong effect on the deformation behavior of the fibers in this region. An excellent discussion by Weigmann and Rebenfeld(*15*) on the role of the disulfide interchange in fiber deformation, particularly at the "turnover" point from the yield to the postyield region, highlights the unique role of the disulfide linkage interchange mechanism which is catalyzed by sulfhydryl groups:

$$R-S-S-R' + R''-SH \longrightarrow R''-S-S-R + R'SH$$

The fiber has a density of 1.304, shows birefringence, and is hygroscopic. The hydroscopicity of wool is one of its most important features, because practically every other property changes with its moisture content. In industrial processing, wools of different histories of wetting, drying, and chemical treatments can show a wide range in moisture regain at the same relative humidity. American standards for moisture regain range from 13–15%, and British standards range from 16–19%.

Single keratin fibers such as wool and hair have been examined in various humidities for changes in mechanical properties. The changes in the complete load-extension curve, that is, in the Hookean yield and postyield regions have been studied while varying the rate of loading, fiber pretreatment, and wool sources. The important role that water plays in modifying the fiber structure is reflected in variations of the mechanical properties observed, as in Fig. 3.

Coulson(*16*) showed that water molecules in keratins are able to form up to four associations in hydrogen bonding, suggesting a continuous molecular network with hydrogen-bonded water serving as crosslinks. Water also acts as a swelling agent and plasticizer reducing interchain interaction. At very high rates of extension(*17*) the Hookean modulus of the keratin–water network is unaffected by the presence of

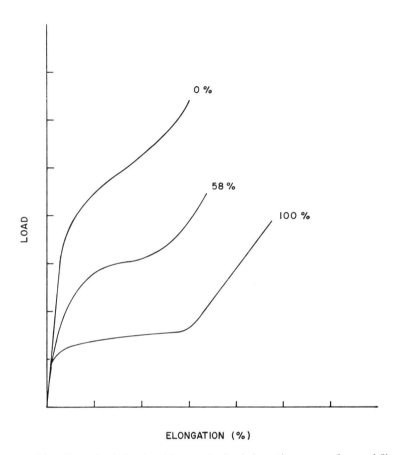

LOAD

ELONGATION (%)

FIG. 3. The effect of relative humidity on the load-elongation curve of a wool fiber.

water. This is because there is insufficient time for the water molecules
to vacate and for the keratin chains to move into the space left vacant.
Thus, there is no plasticizing action under this condition. If the water
had acted mainly as a swelling agent, then the Hookean modulus would
be affected whether the extension times were 1 μsec or 1 sec. For the
movement of water molecules out of a network to be important,
measurements over a long period of time are necessary for creating
vacant sites and allowing chain movement. Movement of water molecules for plasticizing action should be a function both of the number
of water molecules present and the binding energy of these molecules
in the network.

The Hookean region of the stress–strain curve shows a complicated
relationship with relative humidity. And at the turnover point into the

yield region, the stress is nearly a linear function of the relative humidity and it decreases with increasing humidity. Thus, Feughelman and Robinson(*18*) suggest that this turnover, the point at which the ordered keratin of the microfibrils in wool begin to unfold, occurs at the same fiber length. This is better understood by remembering that the reduction in length of a wool fiber near 0% relative humidity as compared with the length at 100% relative humidity involves a compression or distortion in the microfibrils, or both.

The changes that occur in the water sorption properties at various stages during the conversion of raw wool to finished fabric have recieved limited evaluation. Nitschke(*19*) has summarized the pertinent literature, although most data are concerned with the rate of change of water content under nonequilibrium conditions. This, naturally, is governed by the physical state of the assembly of fibers, whether they be loose wool, top, packaged yarn, or woven fabric.

The effect of uv light(*20*) on wool usually results in yellowing. On the sheep, uv degradation appears to involve the decomposition of cystine–disulfide bonds in the wool. The tip of the fiber particularly has been shown to break down chemically causing more hydrophilicity. Such a fiber is more reactive, generally, and would show greater dye absorbance as well as more nonuniformity of absorption. Such wool is often referred to as "tippy" wool or the term "tippy" dyeing might be used. More recently, Asquith and Brooke(*21*) approached the problem of wool yellowing by fractionating wool keratin and exposing the fractions to uv radiation at different pH values. It tentatively appears that cystine and tyrosine are involved in the yellowing process.

IV. Chemical Properties

A. HEAT (AND WATER)

As one would expect from the chemical and physical structure of wool, heat is bound to have an effect on these structures. Earlier work shows that it is not possible to dry wool completely in air at temperatures where decomposition of the wool does not occur. It is also not unexpected that decomposition occurs faster in the presence of water than its absence (or near absence) at any given temperature.

A review of the action of heat on wool can be found in the work of Bell and coworkers(*22*), and Manefee and Yee(*23*). Because of the variety of temperatures employed in this past work, as well as the problem of "dry" wool, there is some ambiguity in comparing the data of different workers. Oxygen was seldom excluded so a number of secondary reactions could occur, thus complicating the physical behavior of the wool fiber. The application of differential thermal analysis (DTA) to a protein fiber was first reported by Schwenker and Dusenbury[24]. At a heating rate of 10°C/min, the major reaction endotherms observed were 112 and 258°C. This first endotherm observed varies[25] depending on the amount of water present, heating rate, level of disulfide reduction, and type of wool. It is probably composed of a normal drying of ordinary water as well as the vaporization of bound water, which has been shown to occur at 130–150°C depending on experimental conditions. The second endotherm at 258°C is in a general pyrolysis area[26] and is probably no more than that[23]. Transition temperatures have been reported in the literature in the range of 200–250°C when dried fibers are heated in vacuum and interpreted to represent a melting of the wool protein, or reactions such as crosslinking or disulfide bond cleavage.

In 1950 Zahn[27] reported the fission of hydrogen bonds, disulfide bonds of cystyl residues, and C–N bonds of acid amide groups of asparaginyl and glutaminyl residues in wool when immersed in water at 100°C. It was also discovered that lanthionyl residues were formed in wool during treatment with water, especially at high pH values. It had already been observed that there was a strong correlation between the number of disulfide bonds degraded and the wet breaking load or abrasion resistance of wool after treatment in water at 100°C [28, 29]. Chemically, it appears that during treatment of wool with water at 60–100°C(pH 5.8) the predominant change is the initial conversion of cystyl residues to lanthionyl residues[30]. The treatment temperature and time determine the extent of conversion, but even after prolonged treatment only about one half of the cystyl residues are converted to lanthionyl residues. This may be due either to reactivities of the various cystyl residues, their location, or perhaps to the mechanics of conversion.

The formation of a lanthionyl group (a monosulfide) requires the loss of one sulfur, whose chemical form has aroused considerable interest. Apparently this sulfur is not in the form of thiol groups or as absorbed sulfide and is called "mobile sulfur." It has been suggested that the reactive intermediates such as hydrosulfide, trisulfide, or sulfenic acid groups could be responsible for this sulfur. Aside from minor changes in tyrosyl or lysyl, or glutaminyl and asparaginyl residues, slow hydrolysis of the peptide bonds occur, producing peptides which dialyze out during a boiling water treatment. The keratin's heterogeneity of amino acids suggests a certain selectivity of fission points, such as peptide bonds adjacent to aspartyl residues (*31*). This important area of research should be followed with interest, for not only are the physical properties of wool affected by water at elevated temperatures, but a study of the water soluble peptides elucidates the amino acid arrangements in the wool fiber, especially in the matrix.

Haley and Snaith(*32*) have examined by DTA, wool in vacuo and with various amounts of water. In all cases, it was observed that a phase transition endotherm and often an endotherm doublet occurred. For a Merino wool, the T_m was 142°C when the heating rate was 38°C/min in a sealed tube and when the water present was greater than 80% of the dry-wool weight. Horio and coworkers(*33*) had found prior to this that at 140°C, wool heated in sealed tubes showed an increasing loss of cystine as the amount of water was increased in the sealed sample. In their DTA work, the sample after fusion showed no x-ray reflections other than a diffuse ring at 9.8 Å and the usual halo at 4.15 Å. Relative entropies of transition were determined from the DTA curves and showed a correlation with the x-ray results.

If wool fibers are extended in steam for at least 30 min they become "permanently set," and subsequent equivalent thermal treatments in water will not destabilize them. Little or no set is produced by heating dry wool fibers unless the temperature exceeds 200°C(*32*). This set or stabilization of an extended state occurs through disulfide bond breakdown and subsequent reformation. Permanent set may also be obtained by chemical methods, as in todays chemical curling and waving of human hair. However, if wool fibers are steamed for only a few minutes, a reduction of the fiber length is observed. This phenomenon is called supercontraction and may be described as a melting of the molecular structure involving extensive rearrangement of hydrogen bonds and disulfide linkages. This contraction in length may also be obtained by treating wool with phenol, sodium sulfide, formamide, and other chemicals capable of breaking hydrogen bonds.

The chemical changes that occur in wool under "wet" heat-setting conditions show once again the conversion of cystyl residues to lanthionyl residues and, in addition, probably to lysino–alanyl residues. Also, there is degradation of some lysyl and aspartyl residues at 150°C(*34*), as well as probable losses of prolyl, threonyl, seryl, isolenyl, phenylalanyl, and asparaginyl or glutaminyl residues. Peptides are formed by slow hydrolysis under wet heat-setting conditions and are leached from the wool.

B. EFFECTS OF ACID AND ALKALI

The physical properties of wool are changed by treatment with acids or bases. The magnitude of these changes depend on the kind of wool, the type of acid or alkaline reagent, its strength, the temperature during exposure, the time of exposure, and, ultimately, on how thoroughly and carefully the acid or alkali is removed after the initial exposure.

Aspartic and glutamic acids, as well as serine, may be freed by partial hydrolysis, while most of the wool protein remains unhydrolyzed. Tryptophan can be almost completely destroyed and there can be losses in cystine and threonine. This degradation of wool usually results in loss of tensile strength and the fiber's sensitivity to acid increases if the cystine has been converted to cystic acid, thus removing some —S—S— crosslinks. Dobozy(35) treated wool with 0.1, 0.01, 0.001 N H_2SO_4 for 30, 60, 120, and 180 min at the boil. After thorough washing and electrodialysis to remove all sulfate, the dried samples showed weight losses of less than 1% except for those samples treated with 0.1 N H_2SO_4 at 60 min. Exposure of these samples to various relative humidities showed increased H_2O absorption. At low humidities water absorption was highest for the lowest acid concentration, yet at high humidities absorption was highest for the highest acid concentrations.

Carbonization of wool becomes necessary when it is contaminated with cellulosic vegetable material that cannot be removed mechanically. This chemical treatment is usually carried out with 5–6% w/v of sulfuric acid. The wool is scoured, immersed in the acid bath, squeezed to about 8% acid on its dry weight, dried, and baked. Drying and baking temperatures and times can vary widely. Temperatures of 180°F for drying and 210°F for baking are frequently employed for 15–30 min for each temperature. After crushing and removal of the carbonized burrs from the wool, neutralization is again carried out, but about 1–2% of acid usually remains. The fiber can lose 10–15% of its tensile strength and 25–35% of its elongation; however, carbonization at the lower temperatures causes little loss of strength. Also, if a wetting agent is employed these losses are claimed to be reduced considerably. Anionic, cationic, and nonionic surfactants are effective, but a nonionic is best(36). This protective action which wetting agents give has not been explained. Carbonization remains a cleaning process which requires careful control, especially in the drying process where sufficient air movement and control of temperatures are necessary.

Raw wool is usually scoured in alkaline solutions at about 50°C.

This temperature is critical, and if over 50°C the conversion of cystine to lanthionine occurs rapidly (*37, 38*) as shown in Fig. 4.

$$H_2N \qquad\qquad\qquad\qquad NH_2$$
$$CH-CH_2-S-CH_2CH$$
$$HOOC \qquad\qquad\qquad\qquad COOH$$

Lanthionine

The urea–bisulfite solubility of wool decreases as the lanthionine content increases and this relationship provides a convenient method for determining alkali damage in wool, since lanthionine analysis is too complicated for quality control (*39*).

Wool fibers can be set by alkaline treatments, as some lanthionine as well as lysinoalanine stabilizing crosslinks are formed (*40*). It is believed, however, that the mechanism for setting is essentially the

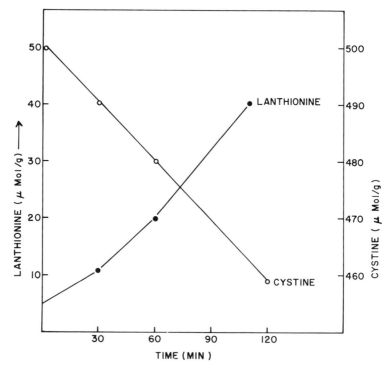

FIG. 4.   Formation of lanthionine and decrease of the cystine content of wool in 0.1 *N* NaHCO₃ solution at 45°C, [Data of Zahn, Blankenburg, and Siepmann (*37*)].

same as in boiling water or bisulfite solutions, that is, *set* is a result of conformational changes in the polypeptide structure which occurs with disulfide bond rearrangement via thiol–disulfide interchange. Other treatments of wool that may reduce the urea–bisulfide solubility are prolonged dry heating, steaming, and heating wet fibers, particularly if the pH is slightly alkaline. This and other solubility tests should be interpreted very carefully. The alkali solubility of wool is an indication of degradation by oxidizing agents and acids. Although it is subject to error, it frequently relates well to the mechanical properties of the fiber(*41*).

C. OXIDATION AND REDUCTION

Many oxidizing agents have been applied to wool in order to bleach the fiber and to prevent felting shrinkage. Hydrogen peroxide is the most important bleaching agent but sodium perborate, percarbonate, and peroxide, persulfate(*42*), and peracetic acid(*43*) and periodic acid(*44*) can be used. Although it is not intended to discuss every oxidizing agent investigated or used industrially, the literature includes ozone(*45*), permangante–sodium chloride(*46*), permonosulfuric acid–sulfate(*47*), bromate–salt(*46*), dichromates and chlorine or chlorine as hypochlorite, chlorite, chlorosulfamic acid, chlorine peroxide(*48*), and dichloroisocyanuric acid(*49*). The chlorine or compounds producing active chlorine are used to produce antifelting or antishrink properties. The chemistry involved is a breakdown of the dithio(disulfide) residues and neighboring groups may affect this reactivity as observed in oxidations with chlorine, permanganate, and peracetic acid. With permanganate, the pH determines the extent and course of the oxidation. While the dithio crosslink is primarily involved with alkaline permanganate only 30%, at most, of the tyrosine groups are oxidized. On the other hand, hydrogen peroxide degrades tyrosine and, under severe conditions, causes fission of peptide bonds as well as the breakdown of serine and threonine and the oxidation of cystine. The peracids are almost specific for breakdown of the cystine crosslinks but they can oxidize methionine and tryptophane also. If prolonged action of chlorine as hypochlorite or chlorine dioxide is used, all of the tyrosine residues are degraded.

The action of oxidizing agents on wool is complex due to the chemical structure of the proteins in the fiber and a variety of chemical reactions are involved. Few mechanisms have been proposed for the observed changes in sulfur and amino acid content and more attention has been given to the resulting properties of the fiber. The chemistry of oxidation has been studied with model compounds(50, 51), but interpretation of the results should be limited. Simply, the complete oxidation of cystine in keratin to give cystic acid may be written as follows:

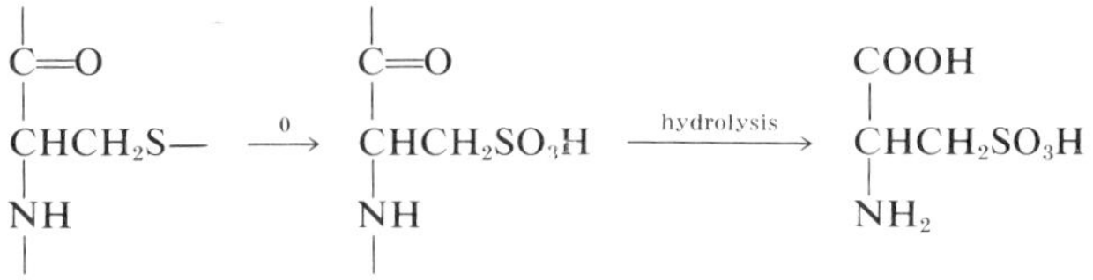

$$
\begin{array}{ccccc}
\overset{|}{C}=O & & \overset{|}{C}=O & & COOH \\
\overset{|}{CHCH_2S-} & \xrightarrow{\ 0\ } & \overset{|}{CHCH_2SO_1H} & \xrightarrow{hydrolysis} & \overset{|}{CHCH_2SO_3H} \\
\overset{|}{NH} & & \overset{|}{NH} & & NH_2 \\
| & & |
\end{array}
$$

However, this reaction is too simple and cannot explain all the data on soluble sulfur compounds in the aqueous phases and the sulfur that remains in the fiber. It is assumed that some of the sulfur remaining in the fiber is inaccessible to the reagents and remains —SS— or as —SH, but this explanation is inadequate. Maclarens, Leach, and Swan(51) studied the intermediate oxidation states of the disulfide groups using small quantities of an oxidizing agent (peracetic acid) and their results suggest disulfide oxide intermediates(52). Large portions of the S—monoxycystyl and probably —SS—dioxycystyl residues can be formed. Also C—S fission appears minor in the peracid oxidations, but this is not necessarily so when other oxidants are used.

Savige and Maclaren(53) have suggested several paths of —S—S— oxidation and propose these as the most likely ones:

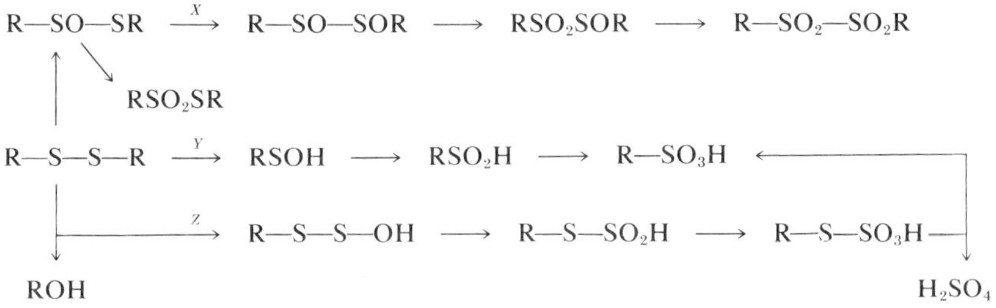

$$
\begin{array}{l}
R-SO-SR \xrightarrow{X} R-SO-SOR \longrightarrow RSO_2SOR \longrightarrow R-SO_2-SO_2R \\
\qquad \uparrow \searrow \\
\qquad | \qquad RSO_2SR \\
R-S-S-R \xrightarrow{Y} RSOH \longrightarrow RSO_2H \longrightarrow R-SO_3H \longleftarrow \\
\qquad | \qquad \xrightarrow{z} R-S-S-OH \longrightarrow R-S-SO_2H \longrightarrow R-S-SO_3H \\
\quad ROH \qquad\qquad\qquad\qquad\qquad\qquad\qquad\qquad\qquad\qquad H_2SO_4
\end{array}
$$

The disulfide oxidation path, regardless of which concept is most likely, will depend on the nature of the chemical environment, solvent, pH, and catalyst. Of the above reactions, it is suggested that the X path would favor anhydrous conditions but that some interplay between X and Y might be possible. In wool, cystyl residues undergo considerable partial oxidation to monoxycystyl residues, and in excess performic acid they have been found to go to sulfonic acid groups.

Reduction reactions in wool refer to the disulfide groups which produce two thiol groups. Such a reaction may be accomplished by treatment with thioglycolic acid (54).

$$
\begin{array}{l}
-CH \\
\ \ |\ \ \\
CH_2 \\
\ \ |\ \ \\
S \quad + \ 2HSCH_2COOH \ \longrightarrow \ 2 \\
\ \ |\ \ \\
S \\
\ \ |\ \ \\
CH_2 \\
\ \ |\ \ \\
-CH
\end{array}
\qquad
\begin{array}{l}
-CH \\
\ \ |\ \ \\
CH_2 \ + \ (-SCH_2COOH)_2 \\
\ \ |\ \ \\
SH
\end{array}
$$

The amount of cystine reduced in wool is dependent on pH and appears to be constant at pH 2–6 with a considerable increase in reduction as pH is further increased. For more than 60% of the cystine to be reduced, repeated treatments have been found necessary in neutral or acid solutions.

Other reducing agents which have received considerable attention include sodium bisulfite and formaldehyde, because sulfur dioxide is a bleaching agent and bisulfite is used as an antichlor aftertreatment with halogens for shrinkproofing. Sodium sulfoxylate ($NaHSO_2$) is another reducing agent widely used on wool. As would be expected, the removal of the disulfide crosslinks reduces the strength of the fiber considerably. However, these thio groups reoxidize readily to the disulfide and almost all of the original strength may be restored.

V. Other Reactions in Finishing

A. INTRODUCTION

The complex structure of a wool fiber presents problems in harvesting, cleaning, and processing. It has scales on its surface which cause one-directional frictional migration producing felting; it has high moisture absorption which causes swelling; and it is highly resilient and can shrink under relaxation conditions. Furthermore, in order to maintain good color and fiber physical properties, cleaning, bleaching, setting, and other treatments must be carefully controlled. Wool fabrics can be made dimensionally stable and smooth drying by proper chemical treatments to control felting shrinkage. To accomplish this, the scales on the fiber can be removed or their edges coated to the fiber surface. Relaxation shrinkage can be controlled by the introduction of crosslinks which are more stable than —S—S— bonds, by grafting, by *in situ* polymerization, or by covering the fibers with a more hydrophobic polymer.

B. ALDEHYDES

Aldehydes have not proven to be useful chemicals for stabilizing wool, although formaldehyde can react with amino, amido, guanidyl, hydroxyl, phenolic, and indole groups and reduce disulfide bonds. It cannot react with two amino groups but it seems to crosslink guanidyl groups as well as an amino group and an amide group(55). The principal advantage observed with formaldehyde is the reduced solubility of the fiber in alkali and, fortunately, this apparent protection does not cause any corresponding change in mechanical behavior. More recent work indicates the formation of crosslinks when "set" fabrics were treated with HCHO because good stabilization of "set" is obtained (56). The exact role of formaldehyde in wool keratin is still being defined(57–59), and these studies will probably aid in the elucidation of fiber structure. The present proposal is that under certain conditions crosslinks are in intimate association with the helical regions(60).

Stabilization of wool is also claimed upon treatment with glutaraldehyde(61) in a rather long reaction cycle requiring additions of sodium carbonate to maintain pH 8. Another method for rendering wool shrink resistant is by treating with a methylol derivative of a mono- or dihydrazide(62). Benzoquinone reacts with wool to form crosslinks, and although it discolors the fiber, some use has been made of its reactivity and molecular size in studying the location of crosslinks. It will react with amino and thiol groups in the following manner as well as polymerize with itself(63). The changes observed in the treated fiber are increased strength and a reduction in felting.

C. MODIFICATION BY OTHER POLYMERS

The properties of wool can be modified (1) by coating its surface with a polymer from solution, (2) by internal and surface deposition of a polymer via interfacial polymerization, (3) by graft polymerization, and (4) by crosslinking a polymer deposited on the fiber's surface.

To apply a polymer from solution to woolen fibers or fabric seemingly involves a pad, squeeze, and dry routine and the property changes vary depending on the properties of the polymer applied and the weight pickup. Such a surface coating could mask the scales and reduce shrinkage due to directional friction effects (felting); it could control fiber swelling; and it could introduce some fiber to fiber bonding. All or each of these would minimize shrinkage. Such coatings would not cause any damage to the fiber but some loss of hand, color, or dyeability might be experienced.

The pretreatment of the wool determines its surface characteristics and this controls the spreading and adhesion of the polymer. The effectiveness and stability of the polymer film is also controlled by the moisture content of the fiber at the time of polymer application and subsequent solvent removal. Polymer migration may be experienced, although this is difficult to detect. It has been proposed(64) that if the critical surface tension of wool is increased from 45 dyn/cm ("normal" wool) to 65–70 dyn/cm by pretreatments, then better spreading and adhesion of polymers will result. The better the polymer spreads and wets the fiber, the closer is the interfacial contact and the better the adhesion. Further study is required for verification of this theory, and consideration should be given to the film strength, flexibility, thermal behavior, and uv stability of the coated polymer.

Polymers which have been deposited on wool to reduce felting shrinkage include: (1) melamines(65), (2) urea–formaldehyde(66), (3) polyamides(67), (4) polyamides with epoxy resins(68), (5) acrylics (69), (6) silicones(70–73), (7) polybutadiene(74–75), (8) polyacrylates (76), and (9) polyamide–epichlorohydrin(77).

The carboxylated butadiene is prepared by treating a polybutadiene with thioglycolic acid, replacing about 5% of the double bonds with —CH_2CHSCH_2COOH. This rubberlike polymer at add-ons of 1.5–2% give good shrinkage resistance (7% versus 63% for untreated fabric). The carboxylated polybutadiene may be treated with ketene and the carboxyl groups converted to an anhydride. The application of this material is slightly more effective, although other similarly modified polydienes may be used. The fabric shows no change in hand and the coating is permanent.

The polyacrylates, studied by Feldtman and McPhee(76), are "soft" and are self-crosslinking when catalyzed with ammonium chloride at 140°C. The application of these polymers offers shrinkage, abrasion, and pilling resistance but soiling problems can be encountered depending on the use of the fabric. In an effort to make this coating more economically feasible, an exhaustion method of application has been evaluated. It consists of a pretreatment with a cationic polymer, polyvinylimidazole disulfate, followed by the polyacrylate dispersion containing a small percentage of an anionic detergent. Exhaustion of a polyamide–epichlorohydrin resin solution may also be accomplished in this manner(77). Additional reactive polymers which have been applied to wool could be described as a polyethylene type(78). Chlorosulfonated polyethylene is effective for shrinkage control but its use requires extensive cure times which are not practical. However, such structures as

$$\left[\begin{array}{c} -[CH_2-CH-(CH_2-CH_2)_9-]_5-CH_2-\overset{\overset{\displaystyle CH_3}{|}}{\underset{\underset{\displaystyle Cl}{\|}}{C}}- \\[2em] \underset{\underset{\displaystyle CH_3}{|}}{\overset{\overset{\displaystyle O}{|}}{C}{=}O} \end{array}\right]_n$$

can be made containing varying amounts of ethylene, methacryloyl-chloride, and vinyl acetate. This polymer is soluble in aromatic petroleum solvents and can be applied in a normal dip, pad, and cure process. These reactive polymers are similar to those in the polyacrylate method which the authors describe as a "phase" boundary limited crosslinking(79). This method involves the application of preformed polymers containing reactive groups. The polymers may then be easily crosslinked forming an ultrathin film on the fiber surface. The method avoids the need for a curing process and requires very low solids pickup to effectively control felting shrinkage. The technique has been extended to include polyesters and polyethers to yield polyurethanes (80). The crosslinking agent used in both cases are diamines such as ethylene-, 1,4-butane-, and hexamethylene-. The crosslinked urethanes give an improvement in most physical properties with good to excellent shrink resistance, although the urethane elastomers themselves have poor solubility in aliphatic hydrocarbons. Also, depending on the elastomer and its isocyanate groups, reaction may occur with active hydrogen atoms in the wool protein(81).

Other polymers which are reactive and have been applied from solution are an ethylenic terpolymer containing acid chloride groups (82); polyethylenimine, epichlorohydrin, and a polyacrylate(83); and nylon 6, 10(84).

D. GRAFTING

Considerable *in situ* polymerization work has been done with vinyl monomers, especially acrylonitrile and the acrylates. These compounds may be polymerized by a free-radical mechanism or in a simple ionic redox system such as hydrogen peroxide–ferrous sulfate(*85*), lithium bromide–potassium persulfate(*86*), or bis(acetylacetonato) copper in trichloroacetic acid–methanol(*87*). Vinyl monomers such as styrene have been polymerized *in situ* after radiation. Preradiation techniques, however, produce diffusion controlled grafting, thus limiting the location of the polymer in the fiber(*88*). When free radicals are first generated in the wool and then treated with the monomer, a rapid decay is noted(*89*). This has been attributed to the faster diffusion of the monomer diluent (water, methanol) which causes the trapped radicals to mutually terminate each other before the monomer reaches them. When the monomer is applied first and then grafted by radiation, different water sorption properties result. These sorption properties, of course, are also influenced by the nature of the grafted polymer. An acrylamide or acrylic acid graft would not show the water sorption effects of a styrene graft(*90*). Stannett and coworkers(*91*) have attempted to determine the location of the grafts and suggest that most of the polymer is located in the matrix.

Other *in situ* polymer reactions which have been considered for shrinkproofing wool include a nylon from bis(p-nitrophenyl) sebacate and hexamethylenediamine(*92*), diisocyanate with a diamine(*93*), and polymers from aziridinyl compounds(*94*).

The shrinkproofing of the wool fiber accomplished by these chemical treatments is described as due to a coating of the scales, "spot welding," reduction of moisture absorption and alteration of elastic properties. However, research on the mechanism of shrink resistance(*95*) in a woolen fabric which had received an interfacial polymerization treatment with (1) a polyamide-6, 10, (2) polyurethane, (3) polyurea, and (4) a polyester still left a mystery. The shrink resistance of the fabrics could not be explained on the basis of the above explanations. The only measurable changes in the treated fibers were an increase in fiber friction in both directions of the fiber and a fairly complete masking of the fiber's scales.

REFERENCES

1. M. Ramanathan, J. Sikorski, and H. J. Woods, Int. Wool Text. Res. Conf., Aug. 1955, F92.

2. M. Horio and T. Kondo, *Text. Res. J.*, **23**, 373 (1955).

3. E. H. Mercer, B. L. Munger, G. E. Rogers, and S. I. Roth, *Nature,* **201**, 367 (1963).

4. E. O. P. Thompson and I. J. O'Donnell, *Aust. J. Biol. Sci.,* **17**, 973 (1964).

5. R. D. B. Fraser, T. P. MacRae, and G. E. Rogers, *Nature,* **193**, 1052 (1962).

6. F. H. C. Crick, *Nature,* **170**, 882 (1952); *Acta Cryst.,* **6**, 689 (1953).

7. R. D. B. Fraser, T. P. MacRae, A. Miller, F. H. C. Stewart, and E. Suzuki, CIRTEL, Paris, 1965, 1-85.

8. L. Pauling and R. B. Corey, *Nature,* **171**, 59 (1953).

9. D. J. Johnson and P. L. Speakman, CIRTEL, Paris 1965, 1-173.

10. E. G. Bendit, *Text. Res. J.,* **38**, 15 (1968).

11. W. G. Crewther, I. J. O'Donnell, J. M. Gillespie, B. S. Harrap, and E. O. P. Thompson, CIRTEL, Paris, Sec. 1, 475 (1965).

12. E. G. Bendit, *Nature,* **179**, 535 (1957).

13. E. G. Bendit, *Text. Res. J.,* **30**, 547 (1960).

14. M. Feughelman, A. R. Haly, and P. Mason, *Nature,* **196**, 957 (1962).

15. H. D. Weigmann and L. Rebenfeld, "The Chemistry of Sulfides," ed., A. V. Tobolsky, Wiley-Interscience, 1968, p. 185.

16. C. A. Coulson, *Research (London),* **10**, 149 (1957).

17. M. Chaikin and N. H. Chamberlain, *J. Text. Inst.,* **46**, 144 (1955).

18. M. Feughelman and M. S. Robinson, *Text. Res. J.,* **37**, 441 (1967).

19. G. Nitschke, *Melliand Textilber.,* **46**, 11 (1965).

20. B. C. M. Doreset, *Text. Mfr.,* **89**, 437 (1963).

21. R. S. Asquith and K. E. Brooke, *J. Soc. Dyers Colour.,* **84**, 159 (1968).

22. J. W. Bell, D. Clegg, and C. S. Whewell, *J. Text. Inst.,* **51**, T1173 (1960).

23. E. Menefee and G. Yee, *Text. Res. J.,* **35**, 801 (1965).

24. R. F. Schwenker and J. H. Dusenbury, *Text. Res. J.,* **30**, 800 (1960).

25. A. R. Haley and J. W. Snaith, *Text. Res. J.,* **37**, 898 (1967).

26. W. D. Felix, M. A. McDowell, and H. Eyring, *Text. Res. J.,* **33**, 465 (1963).

27. H. Zahn, *Melliand Textilber.,* **31**, 481 (1950).

28. F. G. Lennox, *Proc. Int. Wool Textile Res. Conf. Aust.,* **B**, 23 (1955).

29. R. V. Perryman, *J. Soc. Dyers Colour.,* **70**, 83 (1954).

30. B. J. Sweetman, *Text. Res. J.,* **37**, 834 (1967).

31. S. A. Bernhard, A. Berger, J. H. Carter, E. Katchalski, M. Sela, and Y. Shatalin, *J. Am. Chem. Soc.,* **84**, 242 (1962).

32. A. R. Haley and J. W. Snaith, *Text. Res. J.,* **37**, 898 (1967).

33. M. Horio, T. Kondo, K. Sekimoto, and M. Funatsu, CIRTEL, Paris 1965, Section 2, p. 189.

34. B. J. Sweetman, *Text. Res. J.,* **37**, 844 (1967).

35. O. K. Dobozy, *Texil-Praxis,* **19**, 1116 (1964).

36. W. G. Crewther and T. A. Pressley, *Text. Res. J.,* **29**, 482 (1969).

37. H. Zahn, G. Blankenburg, and E. Siepmann, *Textil-Rdsch.,* **18**, 523 (1963).

38. H. Zahn, *Textil-Rdsch.,* **19**, T717 (1960).

39. K. Lees, R. V. Peryman, and F. F. Elsworth, *J. Text. Inst.,* **51**, T717 (1960).

40. J. B. Caldwell, L. M. Dowling, S. J. Leach, and Brian Milligan, *Text. Res. J.,* **34**, 933 (1964).

41. F. Peter, *Magy Textiltech.,* **17**, 352 (1965).

42. A. Kantouch and A. Bendak, *Text. Res. J.,* **37**, 483 (1967).

43. B. J. Sweetman, J. Eager, J. A. Maclaren, and W. E. Savige, CIRTEL, Sec. II, 85 (1965).

44. A. Kantouch and A. Bendak, *Text. Res. J.,* **37**, 772 (1967).

45. W. J. Thorsen and R. Y. Kodani, *Text. Res. J.,* **37**, 975 (1967).

46. J. R. McPhee, *Text. Res. J.,* **30**, 358 (1960).

47. U. S. Patent 2,701,178; 2,739,034, Eric T. Fell.
48. *Wool Science Review*, No. 18 (1960).
49. *Wool Science Review*, No 34 (1968).
50. C. Earland and D. J. Raven, *J. Text. Inst.*, **51**, T678 (1960).
51. J. A. Maclarens, S. J. Leach, and J. M. Swan, *J. Text. Inst.*, **51**, T665 (1960).
52. B. J. Sweetman, Joan Eager, J. A. Maclaren, and W. E. Savige, CIRTEL, Sec. II, 85, 1965, Paris.
53. W. E. Savige and J. A. Maclaren, "The Chemistry of Organic Sulfur Compounds," eds., N. Kharasch and C. Y. Meyers, Pergamon Press, 1966, Vol. 2, p. 370.
54. D. R. Goddard and L. Michaelis, *J. Biol. Chem.*, **106**, 605 (1934); **112**, 361 (1935).
55. H. Fraenkel-Conrat and H. S. Olcott, *J. Am. Chem. Soc.*, **70**, 2673 (1948).
56. J. B. Caldwell, S. J. Leach, B. Milligan, and J. Delmenico, *Text. Res. J.*, **38**, 877 (1968).
57. I. C. Watt and R. Morris, *J. Text. Inst.*, **57**, T425 (1966).
58. R. S. Asquith and D. C. Parkinson, *J. Text. Inst.*, **58**, 83 (1967).
59. I. C. Watt and R. Morris, *J. Text. Inst.*, **58**, 84 (1967).
60. I. C. Watt and R. Morris, *Text. Res. J.*, **38**, 674 (1968).
61. U. S. Patent 3,342,542, USDA.
62. U. S. Patent 3,326,630, Olin Mathieson Chemical Corp.
63. R. C. Ghosh, J. R. Holker, and J. B. Speakman, *Text. Res. J.*, **28**, 112 (1958).
64. H. D. Feldtman and J. R. McPhee, *Text. Res. J.*, **34**, 634 (1964).
65. J. R. Dudley and J. E. Lynn, Symposium on Fibrous Proteins, *J. Soc. Dyers and Colour.*, 215 (1946).
66. P. Alexander, *J. Soc. Dyers and Colour.*, **66**, 349 (1950).
67. D. L. C. Jackson and M. Lipson, *Text. Res. J.*, **21**, 156 (1951).
68. C. E. Pardo and R. A. O'Connell, *Am. Dyest. Rept.*, **47**, 333 (1958).
69. F. H. Steiger, *Am. Dyest. Rept.*, **50**, 97 (1961).
70. W. J. Neish and J. B. Speakman, *Nature*, **156**, 176 (1945).
71. P. Alexander, D. Carter, and C. Earland, *J. Soc. Dyers and Colour.*, **65**, 107 (1949).
72. U. S. Patent 3,345,195, Dow Corning Corp.
73. G. S. Bezruchko, *et al.*, *Tekst. Prom. (Moscow)*, **27**, 54 (1967).
74. E. W. Duck and A. R. Friedl, *Text. Res. J.*, **36**, 724 (1966).
75. British Patent 1,064,660, International Synthetic Rubber Co; British Patent 1,067,903, International Synthetic Rubber Co.
76. H. D. Feldtman and J. R. McPhee, *Text. Res. J.*, **35**, 150 (1965); *ibid.*, **36**, 935 (1966).
77. H. D. Feldtman and J. R. McPhee, *Text. Res. J.*, **34**, 925 (1964).
78. D. E. Remy, R. E. Whitfield, and W. L. Wasley, *Text. Res. J.*, **34**, 939 (1964).
79. R. E. Whitfield, A. G. Pittman, W. L. Wasley, and D. E. Remy, *Text. Res. J.*, **34**, 1105 (1964).
80. R. E. Whitfield, D. E. Remy, and A. G. Pittman, *Text. Res. J.*, **37**, 655 (1967).
81. Chikaaki Sakai and Saburo Komori, *Sen-i Gakkaishi*, **22**, 466, 473 (1966).
82. O. C. Bacon and D. E. Maloney, *Am. Dyest. Rept.*, **56**, 319 (1967).
83. Netherlands Application 6,613,490 (March 28, 1967).
84. H. K. Rouette, H. Zahn, and M. Bahra, *Textilveredlung*, **2**, 474 (1967).
85. S. F. Sadova, and A. A. Konkin, *Zh. Vses. Khim. Obshchest.*, **12**, 596 (1967).
86. Kozo Arai, Michiharu Negishi, and Takayasu Okabe, *Sen-i Gakkaishi*, **23**, 70, 379 (1967); Kozo Arai, Michiharu Negishi, and Takayasu Okabe, *J. Appl. Poly. Sci.*, **12**, 2585 (1968); Michiharu Negishi, Kozo Arai, and Sadayuki Okada, *J. Appl. Poly. Sci.*, **11**, 2427 (1967).

87. W. S. Simpson and W. J. van Pelt, *J. Text. Inst.*, **58**, T316 (1967).

88. D. Campbell, J. L. Williams, and V. Stannett, *Advan. Chem.*, **66**, 221 (1967).

89. D. Campbell, P. Ingram, J. L. Williams, and V. Stannett, *Polymer Letters*, **6**, 409 (1968).

90. J. D. Leeder, A. J. Pratt, and I. C. Watt, *J. Appl. Poly. Sci.*, **11**, 1649 (1967).

91. P. Ingram, J. L. Williams, and V. Stannett, *J. Poly. Sci.*, **6**, A-1, 1895 (1968).

92. H. K. Rouette, *Z. Gesamte Tex. Ind.*, **69**, 325 (1967).

93. U. S. Patent 3,357,785, Merck and Co.

94. P. Alexander, *Melliand Textilber.*, **35**, 3 (1954); G. C. Tesoro and S. Sello, *Text. Res. J.*, **34**, 523 (1964).

95. R. E. Whitfield, W. L. Wasley, R. A. O'Connell, and W. Fong, *Text. Res. J.*, **35**, 575 (1965).

Chapter 5 **POLYAMIDE**

I. Introduction

Polyamide research as initiated by Carothers in 1929 ultimately resulted in the product known today as nylon-6,6. This was the first synthetic polyamide developed that reached technical importance as a fiber. Since then, nylon-6 has also become a commercial product and has assumed an important position among man-made textile fibers. The term nylon-6,6 originated from the fact that the polymer is derived from a six-carbon aliphatic diamine, hexamethylenediamine, $(CH_2)_6(NH_2)_2$ and a six-carbon aliphatic dicarboxylic acid, adipic, $(CH_2)_4(COOH)_2$. The diamine is always named first. Nylon-6, on the other hand, is prepared from a six-carbon ϵ-amino monocarboxylic acid in the form of its lactam:

$$
\begin{array}{l}
CH_2-CH_2-C=O \\
| \qquad\qquad\quad | \\
CH_2 \qquad\qquad | \\
| \qquad\qquad\quad | \\
CH_2-CH_2-NH
\end{array}
$$

Nylon-6,6 has also been categorized as an AABB type polyamide, meaning that it consists of two monomers, hexamethylenediamine having two primary amine groups (AA), and adipic acid having two carboxyl groups (BB). Thus, an AABB-type polyamide consists of two monomers each having difunctionality but each being respectively homogeneous in its functional groups. Nylon-6 would then be an AB type, meaning that the monomer is difunctional but the two reactive sites are different and consist of a reactive nitrogen and a carboxylic carbonyl as in the lactam derived from ϵ-amino caproic acid.

Today, these two nylons, -6,6 and -6, make up more than 98% of the polyamides being converted into fibers. Many thousands of monomers have been used in making polyamides which, in turn, have been evaluated in fiber form. None has shown any important fiber property improvements over the -6,6 and -6 nylons. Several of these aliphatic polyamides are discussed in a later section of this chapter.

II. Nylon-6,6

A. PREPARATION AND STRUCTURE

Nylon-6,6 is prepared by a polycondensation reaction from the dicarboxylic acid, adipic acid, and the diamine, hexamethylene-diamine. The equation for this reaction is best written in the following manner.

Salt preparation

$$HOOC(CH_2)_4COOH + H_2N(CH_2)_6NH_2$$

Adipic acid   Hexamethylenediamine

$$\left[H_3N^+(CH_2)_6N^+H_3 \right] \left[{}^-O-\overset{\overset{\displaystyle O}{\|}}{C}(CH_2)_4\overset{\overset{\displaystyle O}{\|}}{C}-O^- \right]$$

6,6-Salt

Polycondensation

$$\left[H_3N^+(CH_2)_6N^+H_3 \right] \left[{}^-O-\overset{\overset{\displaystyle O}{\|}}{C}(CH_2)_4\overset{\overset{\displaystyle O}{\|}}{C}-O^- \right] \xrightarrow[\text{vac.}]{\Delta}$$

$$H_2N\left[(CH_2)_6NH-\overset{\overset{\displaystyle O}{\|}}{C}(CH_2)_4\overset{\overset{\displaystyle O}{\|}}{C}-NH \right]_n (CH_2)_6NH_2$$

The formation of the intermediate salt is necessary to insure that the components are present in equimolar quantities. This allows the polymerization to continue until a high molecular weight nylon-6,6 is obtained by preventing premature termination of the polymer chain through excess acid or amino end groups. The nylon salt in a 50–60% aqueous solution containing a small quantity of acetic acid (0.5% or less) as a viscosity stabilizer is heated in an autoclave (well purged with nitrogen to prevent discoloration). Heating continues until a pressure of 200–230 psi is obtained and then released. The water is distilled off, and toward the end of water removal a reduced pressure is used. The clear melt is not stable at the high temperatures reached (260–280°C) and so it is quickly extruded, quenched, and chipped.

Before or during the polymerization process, a delustering agent, anatase form of TiO_2, is added to the reaction mass. Nylon-6,6 is a rather polar polymer and the polymerization of the nylon salt begins in an aqueous solution. Thus, a uniform dispersion of small TiO_2 particles can be difficult to achieve, but a proper dispersion without agglomeration is essential in the polymer if it is to yield a uniform, high-quality fiber with a minimum of spinning and drawing problems. A semidull fiber contains about 0.3% TiO_2 and a full-dull fiber contains about 2.0% TiO_2.

The polymer is essentially linear and its molecular weight is above 10,000. The crystal structures have been observed in two forms, α and β, although there is little information on the β form. The α form is described as having the molecules fixed by hydrogen bonds to form sheets, and these sheets pack to form the triclinic cell of the α structure (*1*). The chains have a planar zigzag configuration. There is more recent evidence that in oriented nylon-6,6 two crystalline forms exist which may be described as the extended chain and the folded chain conformations(*2*). Its glass transition temperature is reported as 50°C and its melting point at 250–252°C(*3*). These values, however, are strongly dependent on their method of measurement.

B. FIBER PROPERTIES

When dried, the nylon chips are melt spun into filaments which then require drawing. The entire process may be done on one piece of equipment. The drawing imparts improved orientation and allows the development of crystallinity, particularly if heat is applied to the fiber before or during drawing. The resulting fiber has a high tenacity and, depending on the molecular weight and fiber preparation conditions, which particularly affect orientation, tenacities can range from 3.0 g/den. to 10 g/den. Fiber strength is only slightly affected upon wetting, dropping to 2.6–9 g/den. Elongation of commercial fibers

ranges between 16 and 65%, and they have excellent recovery properties up to about 5% extension. The wide variation in tenacity and
elongation is due principally to variables in molecular weight, polymer
purity, and the spinning and drawing conditions. However, it is often
difficult to specify whether the variations of a particular property are
due primarily to the basic chemical and physical structure of the
polymer, or to the manufacturing process. An example of this would
be the fiber finish and its influential role in flex fatigue or abrasion
resistance.

Specific gravity of nylon-6,6 is 1.14 and its equilibrium moisture
regain at 65% relative humidity is 4.5%. The effect of water has been
recently studied by Howard and Williams(4), who measured fiber creep
as a function of moisture content at various loads and temperatures.
Linear viscoelastic behavior was observed at three stress levels, 0.10,
0.35, 0.50 g/den. Time–temperature superposition could be applied
at the two moisture contents of 1 and 2% studied. The effect of this
moisture was to shift the master creep curve 1.45 decades per 1%
moisture without changing its shape.

The chemical reactivity of nylon-6,6 fibers is dependent on the
amide groups and on the end groups. The fibers are unaffected by
water at normal temperatures but in water at 150°C under pressure
complete hydrolysis can take place. This depolymerization can be
accelerated by alkali. Mineral acids, even in dilute solution, can
bring about hydrolysis. However, the immediate loss in fiber strength
in hydrochloric acid is due not to hydrolysis but to the high solubility
of the fiber in the concentrated acid(5). Oxidizing agents such as
hydrogen peroxide, hypochlorite, and potassium permanganate
will destroy the fiber. A discussion of the oxidation mechanism and the
role of water is included later in this section.

Intensive bleaching is seldom required since nylon-6,6 fibers have
a good white color. Discoloration, however, may result from dry-heat
setting at too high a temperature or a similar process condition. For
safe color removal, sodium chloride or peracetic acid may be used
with sodium pyrophosphate as a buffering agent. Other bleaching
compositions have included sodium chloride with peracetic acid(6),
or solutions of $KHSO_5$, $KHSO_4$, and K_2SO_4 with sodium chloride(7).

Through various industrial applications, the chemical properties of
nylon-6,6 have been characterized over a period of many years. The
acid–base properties, for example, have been measured and isoionic
points and end groups determined via titremetry. In fact, end-group
determinations are useful for measuring number–average molecular
weights up to about 20,000(8). Titration studies on nylon-6,6 and −6

give results which are consistent with the idea that nylon possesses a zwitterion structure at neutral pH with carboxyl groups being protonated on acid titration(9). Nylon-6,6 is degraded via oxidative mechanisms, and a considerable quantity of work has been done on the photooxidation mechanism. This decomposition proceeds through the abstraction of the hydrogen on the carbon alpha to the amide group to form a free radical(10). With oxygen present, this α-carbon radical can react to form a hydroperoxide. When titanium dioxide is used as a delusterant in the fiber, photodegradation is increased above 3000 Å, but only in the presence of oxygen. This then becomes a photooxidation type reaction and it is suggested that TiO_2 acts as a photosensitizer (catalyst).

To date, studies on the oxidation of nylon in the presence of oxygen and water at temperatures below 100°C have been less extensive. Mikolajewski, Swallow, and Webb(11) found that undrawn nylon-6,6 in solutions of hydrogen peroxide resulted in more chain scission than when water and gaseous oxygen were used. Their conclusions were that the carbon atom adjacent to the nitrogen of the N-substituted amide was the point of attack and the degradation products were explained by a free-radical mechanism. Further investigation shows that degradation in oxygenated aqueous systems is pH dependent and reaches a maximum of pH 8.3(12). From this work, using semi-dull fiber, the postulation is that the hydroperoxides formed act as initiators for an autooxidation reaction which may be catalyzed by certain metal ions.

The overall result of oxidation is a decrease in tenacity and elongation with embrittlement and a lower moisture regain. These are not rapid reactions, and uv light absorbers and other protective agents may be incorporated into the fiber to minimize these reactions. Degradation in the absence of air or oxygen at elevated temperatures is most probably related to cleavage of the C–N bond with formation of a double bond and nitrile group(13). The primary degradation may be simply described as follows:

The volatile products from thermal degradation of nylon-6,6 and -6 are water, carbon dioxide, ammonia, hexamethyleneimine, n-hexyl-amine, n-heptylamine, and methylamine. Nylon-6,6 is less stable thermally than nylon-6 and starts to gel (crosslink) under steam at 300°C in about 14 h. Although average holdup time in continuous spinning operations is much shorter than that required for gelation, "dead" zones in the system where molten polymer movement is negligible can result in long retention times and subsequent gel particles causing spinning and drawing defects.

High-energy-radiation sources appear to crosslink the polyamides primarily, although chain scission must also occur quite readily.

C. CHEMICAL MODIFICATION

1. *Crosslinking and Copolymerization*

Nylon-6,6, because of its mono-substituted amide repeat units and its amine end groups, can be crosslinked with formaldehyde. The formaldehyde source is usually paraformaldehyde dissolved in methanol with a very small quantity of sodium hydroxide to aid hydrolysis(14). The final pH of the solution is adjusted to 0.6 with oxalic acid before use. From 15–600 g-eq of crosslinks per 10^6 g of fiber have been obtained. The fibers are insoluble in formic acid and resistant to melting. X-ray examination shows that crosslinking occurs in the amorphous areas. The fibers retain good dyeability, strength, and resilience. The effect of crosslinks on stress relaxation has been studied in detail by Gerasimova and coworkers(15), who followed the resistance to deformation versus time during a protracted postcure. The ultimate number of crosslinks obtained was $270 \times 10^{19}/cm^2$ of fiber.

Nylon-6,6 has also been reacted with diisocyanates and diacid chlorides(16), and the change in physical properties determined. Nylon-6,6 yarn, one highly ordered from a polymer of molecular weight 21,000 and one less ordered from a polymer of molecular weight 14,800, were used. The yarn was treated under conditions of constant length and relaxed. The acid chlorides, although they crosslinked the yarn, also affected the physical properties due to chain scission and reduced interchain forces (hydrogen bonding). However, the diisocyanates increased the tenacity and modulus of the yarn with the greatest changes in properties occuring in the less ordered yarn when held under constant length conditions. Isocyanate functional siloxanes can also improve the tear strength, abrasion resistance, and water resistance of nylon-6,6(17).

Modification of fiber properties can easily be brought about by co-polymerization. Comonomers of recent interest appear to be those which give basic dyeability to the polymer. These include calcium 5-sulfoisophthalic acid, sulfanilic acid, 2,5-aniline–disulfonic acid(*18*), sodium 3,5-bis(methoxycarboxyl) benzene sulfonate(*19*), and 2,6-disulfonates of carboxybenzene(*20*). Cyclohexylphosphonic or cyclo-pentylphosphonic acids(*21*) have been used to give improved acid dyeability. Considerable research(*22*) on aliphatic–aromatic poly-amides has been carried out in order to improve the thermal proper-ties that are associated with flat spotting in tires. However, this approach has led to only limited commercial success. Other modifica-tions include blends of nylon-6,6 or nylon-6 with amino acid co-polymers(*23*), blends of nylon-6 with copolymeric 6,6 containing long-chain amino alcohols and terephthalic acid(*24*) for crimpable fibers, and additions of oleic acid dimer or hydroxy fatty acid or fatty-acid amide for built-in lubricity(*25*).

2. *Grafting*

The chemistry of grafting has been applied to polyamide fibers as to every other textile fiber and has been evaluated as a means of modifying and improving fiber properties. The properties sought range from increased hydrophilicity to increased melting point. When nylon-6,6 is reacted with ethylene oxide to form a hydroxyethyl derivative (*26*), the ethylene oxide is present largely as chemically bonded poly-ethylene oxide chains and produces a flexible material with higher water sorption properties. The "hydroxyethyl nylon" retains the solubility and high-melting behavior of nylon but becomes flexible, hydrophilic, and also shows the second-order transition behavior of the polyethylene oxide chains.

In order to increase the melting point of the fiber, maleic acid and acrylic acid were grafted and a change from 250 to over 350°C(*27*) was obtained. Grafting may be initiated by use of hydrogen peroxide and a water-soluble formaldehyde sulfoxylate salt (Na)(*28*). The acrylic acid graft is converted to the sodium salt by heating the fabric in a sodium carbonate solution and then replacing the sodium ions with calcium ions from a calcium acetate solution. The nylon–acrylic acid graft yields fabrics which have a high resistance to "hole melting." This graft in the calcium salt form has shown promise in flash-resistant fabrics(*29*). Styrene, vinylidene chloride, and methyldodeca(oxy-ethylene) methacrylate have been similarly grafted to nylon-6,6 using heat as the sole initiator(*30*).

Grafting may also be accomplished using high-energy sources either by γ radiation or a Van de Graaf generator. The nylon fabric is made flame resistant by a radiation graft copolymer of vinylidene chloride and vinyl acetate(*31*). The monomers are padded on the fabric and placed in aluminum foil and then irradiated. Improved acid dyeability, reduced static buildup, and increased soil resistance may often be obtained by grafting a vinylpyridine, an unsaturated amine, or a quarternized amine(*32*). Sykes and Thomas(*33*) used 4-vinyl pyridine in their grafting studies. They irradiated nylon-6,6 in air and then heated it in vacuo in the presence of monomer. This produced more grafting than by irradiating the polymer in vacuo followed by treatment with the monomer or by irradiation of the polymer and monomer together.

D. ADDITIVES

1. *Stabilization*

Nylon-6,6 degrades both by heat and uv light, especially in the presence of oxygen. The mechanisms of this degradation have been outlined. Protection against degradation by these energy sources is obtained usually by the addition of certain compounds to the polymer melt and occasionally adequate protection can be achieved by an aftertreatment of the fiber or fabric. For example, uv oxidative protection is claimed when fibers after steaming are treated with a metal salt such as $SnCl_4$ or $AlCl_3$, and the metal is converted to its hydroxide by neutralization with an alkaline aqueous solution(*34*). Copper and manganese salts are well-known antioxidants and may be added as tungstates or molybdates(*35*), although manganous hypophosphite is very frequently used. Another additive combination using a manganous acetate, an organic phosphite, and 1,1,3-tris(2-methyl-4-hydroxy-5-tertbutylphenyl) butane(*36*) gives good protection toward heat, light, and oxidation. The molten polyamide at 290°C showed no gel formation (indicative of degradation) under water vapor after 12 h, and 25% after 16 h. A comparable polymer containing no phosphite or substituted arylalkyl derivative showed 10% gel after 12 h and 61% gel after 16 h.

Ultraviolet absorbers include benzotriazoles(*37*) such as 2(2'-hydroxy-phenyl)-benzotriazole at about 0.25% level. This type of compound with manganous hypophosphite shows a synergistic effect both in light stability and dye lightfastness. Phenols, amines, and urea derivatives have been found to give some protection against heat, light, and oxidation but more recently organic compounds which contain at least one catechol methylene ether group(*38*) have been found more efficient. The use of copper salts and a halogen-containing compound(*39*) in nylon-6,6 appears to be a generally useful system for stabilization; frequently a phosphorus-containing compound is included(*40*). Nylon fabrics have been surface treated with hydroxylated benzophenones and aryl-substituted resorcylates for added protection against uv light(*41*). Less strength loss was experienced when the fabric received an added heat treatment, especially if a higher temperature (350°F versus 225°F) and a shorter heating time are employed. Better protection was also achieved with compounds containing a larger number of hydroxyl groups rather than those containing a larger number of alkoxy groups. Another surface treatment involves the use of an electrically charged aerosol of an antioxidant which can be sprayed onto oppositely charged filaments passing through an electric field of 80,000 V (*42*).

2. *Antistats and Soiling*

Quite frequently the yarn finish will contain materials which perform several functions, such as giving the fiber desired surface lubricity, preventing static or electrical charge buildup, and perhaps plasticizing the fiber. Just as the surface properties are important for controlling charge buildup, either positive or negative, so are they important in preventing soiling and in aiding soil release during laundering. In many instances these factors may go hand in hand with a good antistat preventing pickup of charged dirt particles as well as having enhanced hydrophilic properties that aid in soil release.

Usually, the antistat is applied as a surface treatment, although addition of a polyethylene glycol to the nylon-6,6 flakes before spinning and subsequent irradiation of the yarn has been claimed to give durable antistatic properties(*43*). Many hydrophilic polymers may be used for antistatic finishes and frequently offer some permanency toward laundering. Gardiner and Holmes(*44*) showed moderately effective antistatic action by the application of cellulose acetate in acetone, which was then saponified to yield a cellulosic coating on the fibers. Also reported to be effective is a polyethylene itaconate(*45*), a copolymer of maleic anhydride and methyl vinyl ether with the acid groups in the salt form(*46*), and a polycondensate of an aliphatic dicarboxylic acid, a hydroxypoly(oxyalkylene) and an amino acid or diamine(*47*). Other formulations may include a polysiloxane and a deliquescent salt such as calcium chloride(*48*). Magnesium, calcium, barium, and zinc salts of amphoteric surfactants containing poly-

(oxyethylene) groups are also excellent as internal antistats because of their good compatibility with the polymer(49).

A polymeric carboxylic acid, such as polyacrylic, may be applied by a conventional pad, dry, and cure process(50) to produce effective, permanent, soil-release finishes which also contribute to static reduction. Soil-resistant and antistatic properties may be obtained by treatment of the nylon fabric with a polycaprolactam–ethylene oxide condensate and dimethylolethylene urea with magnesium chloride as the crosslinking catalyst(51). Or a similar formulation may be a hydroxyalkyl derivative of cellulose or starch with the crosslinking resin and catalyst(52). Carpets are claimed to be made soil resistant by treatment with negatively charged polymethyl methacrylate and the sodium salt of polystyrenesulfonate.

III.   Nylon-6

A. PREPARATION AND STRUCTURE

Nylon-6 is made from caprolactam commercially, although it can also be produced from ϵ-aminocaproic acid. In order to produce a good polymer it is essential to have a monomer of high purity and this

$$
\begin{array}{c}
H_2 \\
C \\
H_2C \diagup \quad \diagdown CH_2 \\
| \qquad \qquad \diagup C=O \quad \text{or} \quad CO(CH_2)_5NH \\
H_2C \diagdown \quad \diagup NH \\
C \\
H_2
\end{array}
$$

has been more easily controlled through the use of caprolactam. The polymerization reaction may be written as follows for the primary process:

$$CO(CH_2)_5NH + H_2O \rightleftharpoons HO[CO[(CH_2)_5NH]]_n H$$

The use of small amounts of water to initiate the polymerization is the usual method, although sodium amine salts and even hydrochloric acid have been used. During the polymerization a side reaction occurs resulting in the formation of cyclic oligomers which may be as large as DP 9. Polymerization conditions and the dependency of molecular weight on time and temperature and catalysts have been described by Reimschuessel(53) among others. Acids, such as acetic, benzoic, or adipic, are used as stabilizers for controlling molecular weight, and they also appear to accelerate the reaction. More exotic compounds which may be found in the literature for controlling molecular weight include diamines, tertiary amines, and metal complexes of bis-hydroxy-phenyl sulfones. The polymerization may be carried out batchwise or continuously in a vertical reactor with baffles or perforated plates to aid retention time.

The polymerization reaches an equilibrium when about 10% of the caprolactam is unreacted. It is necessary that this monomer be removed before converting the polymer into a fiber. This improves the spinning performance and avoids fiber processing problems caused by migration of the monomer to the fiber surface. This removal may be accomplished by washing the polymer chips with water, at 90–100°C. The polymer chips are then dried well and melt spun into fibers.

As a dulling agent anatase TiO_2 is used, as in most fibers, and is added in the polymerization process. The rutile form of TiO_2 is more efficient but is seldom used because its hardness can lead to more wear on metal surfaces during processing.

Polycaprolactam is a polycondensation polymer and therefore is an equilibrium polymer with a number-average molecular weight ranging from 14,000 to 20,000 for fiber production. Its melting point is considerably lower than that of nylon-6,6. It is reported to soften around 210°C and show a crystalline melting at 220–225°C. Its second-order transition temperature is approximately 50°C. The specific gravity of unoriented polymer is 1.13.

B. FIBER PROPERTIES

In the spinning and drawing of a nylon-6 fiber there is a wide latitude in fiber properties that may be obtained depending on process conditions such as spinning speed, quench system, draw ratio, drawing speed, temperature of drawing, and so forth. In general, a staple fiber will have a tenacity of 4–5.5 g/den. and a high-tenacity continuous filament will break at 6.5–9 g/den. Elongation of the staple is high, 35–50%, while a high-tenacity filament will have a reduced elongation of 16–20%. The elastic recovery is 100% up to about 2–5% strain. The water absorbency is 2.5–5% and the fiber does lose a small amount of strength when wet. The abrasion resistance is excellent. The electrical resistivity of a dry fiber is on the order of 6×10^{14} Ω-cm and drops to around 10^8 Ω-cm under conditions of high humidity.

The fiber has a complex structure and x-ray diffraction shows several different phases. These have been described as (1) amorphous; (2) α, Bunn's monoclinic; (3) α, paracrystalline monoclinic; (4) γ, pseudo-hexagonal; and (5) β, hexagonal(54).

The final crystalline form in the drawn fiber is usually Bunn's form which is the most stable. However, the processing history, particularly the thermal treatments, will determine the relative amounts of the various phases. In direct spinning where drawing is immediately carried out after extrusion the γ crystalline phase is found. The stable α form places the polymer chains in an extended planar zigzag configuration with the chain antiparallel in order to achieve complete hydrogen bonding. The effects of water and heat during stretching can induce $\gamma \longrightarrow \alpha$ transitions(55). The treatment of a fiber containing the α form with a solution of iodine–potassium iodide and the subsequent removal of the iodine with aqueous sodium thiosulfate regenerates the γ phase(56).

Nylon-6 fibers are soluble in mineral acids, such as hydrochloric and sulfuric, and in organic acids, such as acetic and formic. The fibers will also dissolve in metacresol, trifluoroethanol, and methanol saturated with calcium chloride. Many organic liquids will swell the fiber such as chloroform, methanol, and chloroacetic acid. It is hydro-lyzed in dilute acids but not in concentrated acids. It is little affected by basic compounds but oxidizing agents such as hydrogen peroxide and sodium hypochlorite can cause loss of strength. Reducing agents do not degrade polycaprolactam.

The thermal degradation scheme described in the discussion of nylon-6,6 applies also to nylon-6, although nylon-6 and -6,10 are far more stable thermally. Thermal oxidation, however, results in decreased strength and yellowing. This can be detected before the yellowing is visible by the development of a strong uv absorption band at 2400 Å (57). Oxidized nylon-6 contains double bonds, imide groups, pyrrole groups, and some hydrophilic groups(58). In air, oxidation increases with temperature and becomes significant above 50°C. The mechanism involves the formation of a hydroperoxide on an alpha carbon, followed by decomposition leading to colored products and chain scission.

When degradation is caused by a high-energy source such as radia-tion or by uv light, chain scission and loss of desirable tensile proper-ties will result. However, the steps to degradation may depend on the presence or absence of TiO_2 in the fiber. Without TiO_2, the poly-mer absorbs the energy directly producing free radicals. In the presence of TiO_2, which acts as a photosensitizer, the TiO_2 absorbs the energy and forms metastable Ti_2O_3, losing atomic oxygen(59). By this mechanism the molecular weight and tensile strength are decreased. Prati(*60*) has found Sippel's(*61*) relationships between loss of strength to degree of polymerization to apply in photochemical and photolytic degradation studies.

C. CHEMICAL MODIFICATIONS

1. *Derivatization and Crosslinking*

Polyamide structures are subject to derivatization in much the same manner as wool fibers in that there are substituted amide groups, amino end groups, and carboxyl end groups. In addition, the polymer's low melting point can be increased and its complex solid state structure can be stabilized by chemical methods.

One of the first chemical treatments given to synthetic polyamides is the preparation of N-methylol, N-alkoxymethyl, and N-alkylthiomethyl derivatives(*62*). These reactions are carried out with formaldehyde to yield an N-methylol derivative or with formaldehyde in the presence of an alcohol or mercaptan to yield the N-alkoxymethyl or N-alkylthiomethyl derivatives. When the methoxymethylation reaction is carried out on nylon-6 fibers, paraformaldehyde in methanol is used with a small amount of potassium hydroxide to solublize the aldehyde via hydrolysis. Oxalic acid may be used as a catalyst and the fibers are then treated with this solution(*63*). The reaction occurs in the "amorphous region" of drawn filaments, thus depriving these regions or chain segments of their ability to crystallize. If methoxymethylation is extensive, an increase in melting temperature is obtained. Arakawa and coworkers(*64*) have used this reaction to study the effect of annealing temperatures on melting phenomena in the crystallites. When hydroxymethylation of nylon-6 fibers reaches 40 wt. %, the fibers begin to adhere(*65*) and these have been pressed in a heated mold to form a sheet that has good impact strength, and is microscopically porous. A possible use for this is as a substitute for leather (*66*).

Another application of an N-alkoxymethylation reaction is to induce crimp in a fiber by treatment in a hot alcoholic formaldehyde bath containing an acid catalyst(*67*). The partial substitution of amide hydrogens, as, for example, the addition of ethylene oxide, can bring about a lowering of melting point and a loss of strength—but the elasticity increases. Partial benzoylation(*68*) up to a degree of substitution (DS) of about 0.1 decreases the elongation and the tenacity about

60–80%. These fibers are less soluble in polar solvents but their dimensional stability in thermal treatments is improved. From a DS of 0.1–0.4, fiber strength continues to decrease to about 40% of its original value but ceases to decrease further up to a DS of 0.8, where crimping effects are evident and thickening of the fiber wall, as well as cracks in the skin, are microscopically visible. As would be predicted, solubility in polar solvents has ceased and swelling in nonpolar solvents has increased.

Nylon-6 can be derivatized with isocyanates, such as propyl 3-isocyanate-4-methylphenyl-carbamate(69) or with an epoxy compound such as vinylcyclohexene dioxide(70), and when the reaction is extensive ($\sim$ 42 wt. % increase for the dioxide) the melting point is increased.

Chemical crosslinks may be formed in nylon-6 fibers by means of adipoyl and sebacoyl acid chlorides, hexamethylene and tolylene diisocyanates, cyanuric chloride and S_2Cl_2 as well as by the synthesis of acetoacetate–polyamide derivatives which are capable of crosslinking with diketene(71). In general, the elastic modulus and thermal stability increase regardless of the crosslinking agent. The most effective crosslinking agent is a diisocyanate and, in addition, fibers crosslinked with these compounds either retain or else increase in breaking strength. The fibers treated with S_2Cl_2 had the greatest thermal stability. While diisocyanates are too toxic to be used directly in industry, this toxicity can be reduced by using the diisocyanate adduct formed with phenol, o-chlorophenol, or methanol(72).

Brucks(73) has studied the introduction of disulfide crosslinks in nylon-6 as well as alkylene sulfide crosslinks. Each type of crosslink was found to produce crimping and coiling in the fiber, and this was increased when the crosslinked fiber was subsequently treated with m-cresol. The disulfide linkage is obtained by methoxymethylation of the fiber followed by treatment with an acidic aqueous solution of thiourea and then potassium hydroxide in the following reactions. The alkylene sulfide crosslinks may also be introduced via treatment of the methoxymethyl derivative with thioacetamide to produce a mercaptide and crosslinking with an alkyl dihalide(74). The alkylene sulfide crosslinked fibers show no decrease in their extent of crimping and coiling on prolonged exposure to air or in dilute solutions of hydrogen peroxide as do disulfide-crosslinked fibers. If the nylon-6 fiber has a crystalline structure of predominately the alpha form, it will exhibit a much greater crimping tendency than those which contain both alpha and beta forms in approximately equal amounts.

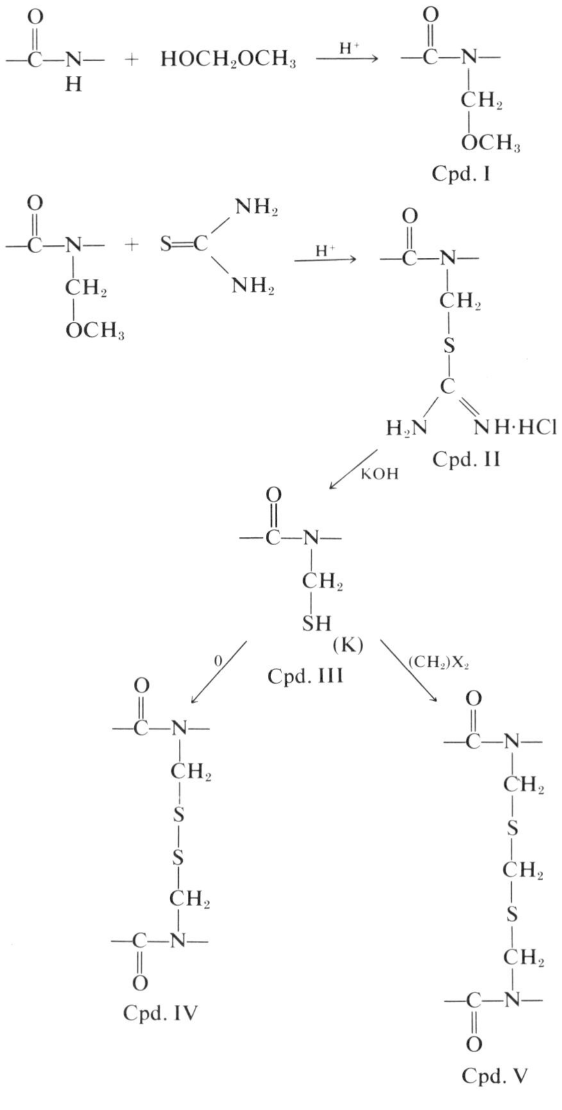

$$-\overset{\overset{\text{O}}{\|}}{\text{C}}-\underset{\text{H}}{\text{N}}- \ + \ \text{HOCH}_2\text{OCH}_3 \ \xrightarrow{\text{H}^+} \ -\overset{\overset{\text{O}}{\|}}{\text{C}}-\text{N}- $$

Cpd. I

Cpd. II

Cpd. III (K)

Cpd. IV

Cpd. V

2. Grafting

Nylon-6 has also been grafted with a variety of vinyl monomers and the property changes of the fibers have been determined. Acrylonitrile grafts have received considerable study because of the possible advantages of handling acrylonitrile in a vapor phase. The reaction may be carried out by radiation or by photochemical initiation(75). Initiation by both methods seemed to show similar rates when the fibers were subjected to free-radical formation prior to exposure to gaseous acrylonitrile. However, when the x-ray irradiated fibers were exposed to air for only a short time, grafting was suppressed(76). When the effect of methanol on the vapor phase grafting of acrylonitrile was investigated, it was found that the methanol had no effect on the permeation of the acrylonitrile into the fiber(77), but later work reported it to aid permeability(78).

Shinohara and Mukoyama(79) have grafted styrene, methyl methacrylate, acrylonitrile, vinyl acetate, methyl acrylate, and ethyl acrylate to nylon-6 by a preirradiation method. With styrene, methyl methacrylate, and acrylonitrile they report an increase in modulus and torsional rigidity with a tensile strength loss of about 50%. Ethyl and methyl acrylate decreased the modulus and torsional rigidity. Vinyl acetate was intermediate between these two groups. The observed property changes, it was suggested, was due to the second-order transition temperatures of the vinyl homopolymers. The second-order transition temperature of polymers from styrene, methyl methacrylate, and acrylonitrile is higher than the acrylates. Dyeability, moisture regain, and heat stability decreased with increasing graft.

It also has been found that methyl methacrylate will polymerize in the presence of nylon-6 and water via thermal (85–95°C) initiation (80). This polymerization can occur within the fiber and the major portion of the homopolymer is not extractable(81). In one example when ammonium persulfate was used as an initiator, a saturated steam treatment was used after application of the initiator and then the monomer(82).

Fluoroolefins have also been grafted to nylon-6. Vinyl fluoride grafted via γ-ray irradiation decreased the glass transition temperature(83). Vinylidene fluoride, trifluorochloroethylene(84), and hexafluoropropylene(85) have been grafted on nylon-6.

D. ADDITIVES

1. *Stabilization*

Nylon-6 is somewhat more stable than -6,6, and the same compounds will contribute to its thermal stability as well as its resistance to uv light. Most synthetic polymers have compounds added to them which will prevent or slow down the thermal and oxidative degradation reactions to which they are subjected. These compounds are usually added during or at the end of the polymerization process. The patent literature on nylon-6 divides this protection into the following categories:

(1) *Heat.* Copper compounds plus halogen compound and/or phosphorus compound.
(2) *Light.* Manganese compounds plus phosphorus compound. $Mn_2P_2O_7$
(3) *Heat and Light.* Combinations of 1 and 2 but predominately manganese and phosphorus compounds.
(4) *Heat, Light, and Oxygen.* Copper, phosphorus, and halogen with some aromaticity in structure.

A few examples of these additives are listed in Table I. The quantities of these compounds used will vary somewhat but usually small amounts, on the order of 0.1–0.5 wt.%, appear to be adequate.

TABLE I

STABILIZATION OF NYLON-6 TOWARD HEAT

Copper addition cpd. of halogenated aliphatic phosphite	Belg. 695,196, Farbenfab Bayer A.G.
Copper complex: example, $Na_3(Cu(CN)_4)$	Ger. 1,247,631, Farbenfab Bayer A.G.
Copper cpd. and quaternary ammonium salt of polyhydriodic acid	Belg. 694,473, Farbenfab Bayer A.G.
Copper cpd., salt of hydrocyanic acid and iodine or phosphorus cpd.	Belg. 695,324, Farbenfab Bayer A.G.
Copper cpd., aromatic amine subst. in nucleus by I_2, Cl_2, or Br_2	Belg. 654,151, Allied Chem. Corp.
Copper cpd., a phosphine and inorganic or organic salt of HI	French 1,491,943, Farbenfab Bayer A.G.
Copper cpd., a phosphine and inorganic or organic salt of HI	British 1,084,699, Farbenfab Bayer A.G.
Copper cpd., $Cu_x(R_1—NHCONH—R_2—NH—R_3)I_y$	Belg. 698,702, Veb Leuna-Werke "Walter Ulbright"
Organic phosphoric acid derivative or org. phosphite	Neth. Appl. 6,601,486, Monsanto Chem. Co.

STABILIZATION OF NYLON-6 TOWARD LIGHT

Manganese cpds. as orthophosphates, pyrophosphates and complex phosphates	U.S. 3,352,821, I.C.I. Ltd.
Manganese phthalate and hypophosphorous acid or metal salt	British 1,088,905, SNIA Viscosa
Diacetyl-benz-sulfhydroxamic acid	Belg. 622,700, Vereinigte Glanzstoff
Sodium borohydride	Belg. 622,701, Vereinigte Glanzstoff
Manganese salt and H_3PO_2	British 968,770, Allied Chem. Corp.

TABLE 1 (Contd.)

STABILIZATION OF NYLON-6 TOWARD LIGHT

Manganese pyrophosphate	Jap. 23094(67), Ashai Chem.
Manganese dodecylbenzenesulfonate	Jap. 7706(67), Kanegafuchi Spin.
Manganese salt of β-napthalene-sulfonic acid	Jap. 01,502(68), Toyo Rayon Co.

STABILIZATION OF NYLON-6 TOWARD HEAT AND LIGHT

Manganese cpd., an amide of phosphoric acid, polyphosphoric or salt of hypophorus	French 1,462,128, British Celanese Ltd.
Manganese salts of phosphoric acid esters	French 1,464,681, British Celanese Ltd.
Cupric pyrophosphate	French 1,464,645, Firestone Tire and Rubber Co.
Manganese salts and halonaphthalene derivatives	Japan 08,617(68), Toyo Rayon Co.
Manganese salts and amine hydrochloride	Japan 08,616(68), Toyo Rayon Co.
Copper pyrophosphate	U.S. 3,425,986, Firestone Tire and Rubber Co.
Manganese pyrophosphate	
Copper stearate, acetate	

STABILIZATION OF NYLON-6 TOWARD HEAT, LIGHT, AND OXYGEN

Copper salt and phosphorus trihalide or oxyhalide	German 1,152,816, Badische Anilin and Soda Fabrik
	Japan 23097(67), Teijin Ltd.
Iodine- and/or bromine-substituted phenols	U.S. 3,341,492, Celanese Corp.
PbCl$_2$, manganese salt, water soluble alkali metal halide	Japan 08,615(68), E. Akazuka
Copper dialkyl phosphate in a triaryl phosphite	Japan 01,503(68), Toyo Spin. Co.
Copper salt or org. copper complex such as 8-hydroxyquinoline, EDTA	Japan 01,504(68), Toyo Spin. Co.

2. *Antistats*

The problems in static buildup on nylon-6 fibers are similar to nylon-6,6 and are controllable in processing the fiber by application of finishes. Mineral oils and fatty acids or their esters are still important base components in lubricant and antistat finishes. Polyols, organic phosphates, amines, and silicate esters are a few of the ingredients which can be found in finish formulations. Water is always an important base because of its role in electrical conductivity. For addition to the polymer melt, metal containing compounds of tin and germanium has been claimed(86) as effective antistats as well as sulfonated and phosphated derivatives of ethylene oxide and monoalkyl phenol adducts(87).

IV. Other Nylons

We have divided the aliphatic polyamides into the AB and the AABB type. Several examples of the AB type that are of interest will be discussed briefly. The aromatic polyamides are also included here because their stability at high temperatures make them very useful fibers in industrial and military areas.

Nylon-3 is made from β-propiolactam but is not attractive commercially. However, substituted β-propiolactams have been examined and one, poly (4,4-dimethylpropiolactam), is very stable to oxidative thermal degradation(88). Nylon-4, poly(pyrrolidone) is obtained from polymerization of pyrrolidone with a reasonable reaction rate using an anionic initiator and an acylating agent. Melt spinning is not possible as it decomposes with reversion to monomer at 265°C. Its fibers are reported to have a high moisture absorption and to be similar to cotton over the whole relative humidity range.

Nylon-7, poly (ω-enanthamide), is more conveniently prepared from 7-aminoheptanoic acid rather than the enantholactam. Its melting point is 223°C and it has a lower water absorption than nylon-6(89). Nylon-8, poly(capryllactam), is thermally stable but its low melting point (200°C) and high monomer cost do not make it a useful fiber.

Nylon-9, poly (ω-perlargonamide), is prepared by melt condensation of ω-amino pelargonic acid at temperatures above 230°C. This polymer has not been evaluated extensively as a fiber because of relatively high cost. Also nylon-10, poly (aminodecanoic acid), involves difficulty in monomer preparation and is a low-melting polymer (188°C). Nylon-11, poly (ω-undecaneamide), has limited commercial use as it is very hydrophobic, has very good electrical insulating properties, and melts at 185°C.

Other polyamides include the AABB type such as poly(hexamethylene terephthalamide), nylon-6,T. It has a high melting point, 370°C, and has been carefully evaluated for fibers(90). Block copolyamides have been studied(*91*) as well as polyamides that are totally aromatic (*92*).

The present commercial aromatic polyamide (Nomex) is probably poly-m-phenylene isophthalamide(*93*). This fiber has excellent long-term thermal stability at temperatures between 200–300°C, retaining its physical and electrical properties at these elevated temperatures. In general, it has good resistance to chemicals and solvents. Concentrated acids such as hydrochloric and sulfuric, particularly above room temperature, will result in a loss of strength. Nomex has resistance to ionizing radiation that is several times better than nylon-6,6. Its moisture regain value is 5% which is somewhat higher than that of nylon-6,6 also.

REFERENCES

1. R. Brill, *Z. Phys. Chem., Abt. B*, **53**, 61 (1943); C. W. Bunn and E. V. Garner, *Proc. Royal Soc.*, **189**, 39 (1947).
2. P. E. Dismore and W. O. Stratton, *J. Poly. Sci., Part C*, **13**, 133 (1966).
3. F. J. Hybart and J. D. Platt, *J. Appl. Poly. Sci.*, **11**, 1449 (1967).
4. W. H. Howard and M. L. Williams, *Text. Res. J.*, **36**, 691 (1966).
5. G. Johansen and L. H. Reyerson, *J. Phys. Chem.*, **64**, 1959 (1960).
6. British Patent 1,148,374, Precision Processes, Ltd.
7. German Patent 1,269,992, DuPont; British Patent 893,440, DuPont.
8. H. Hopff, *"Man-made Fibers, Science and Technology,"* H. F. Mark, S. M. Atlas, and E. Cernia, eds., Wiley-Interscience, New York, 1968, p. 194.
9. A. R. Mathieson, C. S. Whewell, and P. E. Williams, *J. Appl. Poly. Sci.*, **8**, 2009 (1964).
10. A. Anton, *J. Appl. Poly. Sci.*, **9**, 1631 (1965).
11. E. Mikolajewski, J. E. Swallow, and M. W. Webb, *J. Appl. Poly. Sci.*, **8**, 2067 (1964).
12. R. N. Vachon, L. Rebenfeld, and H. S. Taylor, *Text. Res. J.*, **38**, 716 (1968).
13. "High Temperature Resistance and Thermal Degradation of Polymers," Soc. Chem. Ind. London Monogr. 13, 1961.
14. U.S. Patent 3,294,755, DuPont.
15. L. S. Gerasimova, A. B. Pakshver, and A. Kh. Khakimovo, *Mekh. Polim.*, **1968** (1) 53.
16. E. Perry and J. Savory, *J. Poly. Sci., Part C*, No. 24, 89 (1967); *J. Appl. Poly. Sci.*, **11**, 2473, 2485 (1967).
17. U.S. Patent 3,485,661, Dow Corning Corp.
18. F. M. Mandrosova, M. N. Bogdanov, T. V. Kravchenko, and G. V. Scheglova, *Khim. Volokna*, **1968** (5), 14.
19. U.S. Patent 3,389,549, DuPont.
20. German Patent 1,801,366 ICI, Ltd.
21. French Patent 1,512,185, Imperial Chem. Ind.
22. B. S. Sprague and R. W. Singleton, *Text. Res. J.*, **35**, 999 (1965).
23. U.S. Patent 3,419,636, Allied Chemical Corp.
24. British Patent 1,140,479, Snia Viscosa.
25. U.S. Patent 3,329,633, Monsanto Co.; U.S. Patent 3,342,762, Monsanto Co.

26. H. C. Haas, S. G. Cohen, A. C. Oglesby, and E. R. Karlin, *J. Poly. Sci.*, **15**, 427 (1955).
27. E. E. Magat, I. K. Miller, D. Tanner, and J. Zimmerman, *J. Poly. Sci., Part C*, No. 4, 615 (1964).
28. U.S. Patent 3,394,985, DuPont.
29. E. M. Healy and B. A. Natsios, U.S. Clearinghouse Fed. Sci. Inform., AD 650236, 1966 [*U.S. Govt. Res. Develop. Rep.* **67** (11), 95 (1967)].
30. U.S. Patent 3,313,591, DuPont.
31. Netherlands Application 6,606,050, DuPont.
32. U.S. Patent 3,424,820, DuPont.
33. J. A. Sykes and J. K. Thomas, *J. Poly. Sci.*, **55**, 721 (1961).
34. British Patent 1,085,408 Japan Rayon Co.
35. Netherlands Application 6,601,754, I.C.I. Ltd.
36. French Patent 1,546,069, I.C.I. Ltd.; S. African Patent 6,700,978, I.C.I. Ltd.
37. B. C. M. Dorset, *Text. Mfr.*, **93**, 70 (1967).
38. British Patent 1,036,414, Badische Anilin und Soda Fabrik A.G.
39. U. S. Patent 3,359,235, Monsanto Co.; British Patent 1,131,933, Celanese Corp.; French Patent 1,498,023 Badische Anilin- und Soda-Fabrik A.G.
40. Netherlands Application 6,601,486, Monsanto Co.
41. U.S. Patent 3,320,207, American Cyanamid Co.
42. British Patent 1,124,903, Veb Chemiefaserkombinat Wilhelm-Pieck-Stadt Guben.
43. U.S. Patent 3,329,557, DuPont.
44. D. Gardiner and F. H. Holmes, *J. Soc. Dyers Colour.*, **83**, 43 (1967).
45. British Patent 1,143,944, I.C.I. Ltd.
46. S. A. Heap and F. H. Holmes, *J. Soc. Dyers Colour.*, **83**, 12 (1967).
47. French Patent 1,469,254, I.C.I. Ltd.; British Patent 1,148,065.
48. U.S. Patent 3,423,314, Dow Corning Corp.; French Patent 1,561,732, Dow Corning Corp.
49. Hideo Marumo, Makoto Takai, Minoru Saito, and Seigo Oya, *Kogyo Kagaku Zasshi*, **72** (4), 940 (1969).
50. *Textile Month*, 56 (Feb. 1969), p. 56.
51. Netherlands Application 6,505,814, I.C.I. Ltd.
52. U.S. Patent 3,380,850, I.C.I. Ltd.
53. H. K. Reimschuessel, *J. Poly. Sci.*, **41**, 457 (1959).
54. D. R. Holmes, C. W. Bunn, and D. J. Smith, *J. Poly. Sci.*, **17**, 159 (1955); A. Reichle and A. Prietzschk, *Angew. Chem.*, **74**, 562 (1962); L. G. Roldan and H. S. Kaufman, *J. Poly. Sci. Part B*, **1**, 603 (1963); W. Ruland, *Polymer*, **5**, 89 (1964); L. G. Roldan, F. Rahl, and A. R. Paterson, *J. Poly. Sci., Part C*, **8**, 145 (1965).
55. K. Miyasaka and K. Makishima, *Kobunshi Kagaku*, **23** (260), 870 (1966); K. Miyasaka and K. Ishikawa, *J. Poly. Sci., Part A-2*, **6**, 1317 (1966).
56. H. Arimoto, *J. Poly. Sci., Part A*, **2**, 2283 (1964).
57. W. Sbrolli, T. Capaccioli, and E. Bertotti, *Chim. Ind. (Milan)*, **42**, 37 (1960).
58. W. Sbrolli and T. Capaccioli, *Chim. Ind. (Milan)*, **42**, 1325 (1960).
59. W. A. Weyl and T. Förland, *Ind. Eng. Chem.*, **42** (2), 259 (1960).
60. G. Prati, *Ann. Chim. (Rome)*, **48**, 15 (1958).
61. A. Sippel, *Melliand Textilber.*, **33**, 645 (1952).
62. T. L. Cairns, H. D. Foster, A. W. Larchar, A. K. Schneider, and R. S. Schreiber, *J. Am. Chem. Soc.*, **71**, 651 (1949).
63. T. Arakawa, F. Nagatoshi, and N. Arai, *J. Poly. Sci., Part B*, **6**, 513 (1968).
64. T. Arakawa, F. Nagatoshi, and N. Arai, *J. Poly. Sci., Part B*, **7**, 115 (1969).

65. T. Takeshita, S. Maeda, and M. Sasaoka, *Chem. High Polym. (Japan)*, 26, 209 (1969).
66. T. Takeshita, S. Maeda, and A. Nakagawa, *Chem. High Polym. (Japan)*, 26, 217 (1969).
67. U.S. Patent 3,370,912, Heberlein Pat. Corp.
68. G. von Hornuff and M. Jansch, *Melliand Textilber.*, 45, 768 (1964).
69. S. S. Vladyko, A. A. Kachan, L. N. Korsakova, and V. G. Sinyavskii, *Sin. Fiz.-Khim. Poliuretanov*, 1967, 45.
70. British Patent 1,138,053, Soo Valley Co.
71. A. A. Konkin, G. A. Gabrielyan, and O. G. Smolnikova, *Chem. Prum.*, 17 (11), 601 (1967); A. A. Konkin, *Chemiefasern Text.-Anwendungstech.*, 19 (2), 121 (1969); R. W. Moncrieff, *Tex. Mfr.*, 92, 58 (1966).
72. T. B. Bezhuashvili, D. A. Predvoditelev, and A. A. Konkin, *Khim. Volokna*. 1969 (1), 36.
73. S. D. Bruck, *J. Res. Nat. Bur. Stand.*, 65A, 489 (1961); U.S. Patent 3,331,656, U.S. Dept. Commerce.
74. S. D. Bruck, *J. Res. Nat. Bur. Stand.*, 66A, 77 (1962).
75. A. A. Kachan, L. L. Chervyatsova, K. A. Kornyev, E. F. Mertvichenko, and N. P. Gnyp, *J. Poly. Sci., Part C*, No. 16, 3033 (1967).
76. A. A. Kachan, *Vysokomol. Soedin*, 8, 2144 (1966).
77. E. F. Mertvichenko, A. A. Kachan, V. A. Vonsyatskii, and A. M. Kalinichenko, *Vysokomol. Soedin., Ser. A*, 9, 1382 (1967).
78. A. A. Kachan and E. F. Mertvichenko, *Vysokomol. Soedin., Ser. A*, 9, 1424 (1967).
79. Y. Shinohara and E. Mukoyama, *Sen-i-Gakkaishi*, 18, 480 (1962).
80. Sadao Hayashi, *Kogyo Kagaku Zasshi*, 71, 1064 (1968).
81. Sadao Hayashi and M. Imoto, *Angew. Makromol. Chem.*, 1969, 6, 46.
82. Japanese Patent (68) 25,975, Toyo Rayon Co.
83. Kh. U. Usmanov, A. A. Yul'chibaev, and T. Sirlibaev, *Khim. Volokna*, 1967 (6) 32.
84. A. I. Kurilenko, E. P. Danilov, and U. L. Karpov, *Vysokomol. Soedin., Ser. A*, 9, 2362 (1967).
85. A. I. Kurilenko, I. G. Nikulina, and E. P. Danilov, *Vysokomol. Soedin. Ser. A*, 9, 2376 (1967).
86. U.S. Patent 2,924,586, Ver. Glanzstoff-Fabriken.
87. Japanese Patent 22160 (67), Toyo Rayon Co.
88. R. Graf, G. Lohaus, K. Börner, E. Schmidt, and H. Bestian, *Angew. Chem., Intern. Ed. Engl.*, 1, 481 (1962).
89. U.S. Patent 2,241,321, I.G. Farbenindustrie.
90. B. S. Sprague and R. W. Singleton, *Text. Res. J.*, 35, 999 (1965).
91. British Patent 918,637, DuPont.
92. L. K. McCune, *Text. Res. J.*, 32, 762 (1962).
93. W. B. Black and J. Preston, "Fiber-forming Aromatic Polyamides," *Manmade Fibers*, eds., H. F. Mark, S. M. Atlas, and E. Cernia, John Wiley and Sons, 1968, Vol. 2, p. 297.

Chapter 6 ACRYLIC FIBERS

I. Introduction

An acrylic fiber is defined as "a manufactured fiber in which the fiber-forming substance is any long-chain synthetic polymer composed of at least 85% by weight of acrylonitrile units ($-CH_2-CH(CN)-$)." A more recent modification are the modacrylic fibers which are comprised of less than 85% but at least 35% by weight of acrylonitrile. These definitions are generally accepted and are included in the Textile Fiber Product Identification Act (1960). However, confusion may still occur as the less recent literature often refers to the modacrylics as modified acrylics or even vinyl fibers. Acrylic fiber production has more than quadrupled in the past 20 years and the 1970 U.S. fiber production is over 530 million lb(1). The fiber is used in a variety of knits, carpets, blankets, manmade furs, and, more recently, in the production of carbon and graphite fibers by combustion.

II. Structure

A. CHEMICAL

Acrylic fibers have a chemical structure consisting essentially of the repeating unit, ($-CH_2-CH(CN)-)_n$, with up to 15% by weight of the polymer consisting of one or two other monomeric units. As comonomers, vinyl acetate and an acrylate or methacrylate ester is used in order to vary the properties of the polymer for both ease of processing into a fiber and for improved fiber properties. Many other vinyl compounds are listed in the literature(2), though many of these have not become commercially useful. For further improvement, particularly dyeability, a small quantity of even a third monomer may be used. Such a tripolymer can be designed to improve its affinity either to acid or basic dyestuffs. The quantities of these minor monomer constituents are important because of their high cost and the often deleterious effect which they may have on the polymerization rate and on the properties of the fiber.

B. Physical

The molecular weight of polyacrylonitrile, as determined by light scattering, seems to vary between 80,000 and 175,000(3). The effect of molecular weight on spinnability has been studied most recently by Paul(4), and Potalovskaya(5). Paul, using an acrylonitrilevinyl acetate (7.7%) copolymer of molecular weight 115,000, found that broken filaments were due to tension near the spinneret face where the filament was partially coagulated. The spinning solution elasticity contributed to this through the Barus effect. Polatovskaya and co-workers using polyacrylonitriles varying in molecular weights from 25,000 to 100,000 found the porosity of the fiber decreasing, the modulus increasing and the fatigue resistance improving with increasing molecular weight. The acrylic polymers are considered stiff molecules due to intramolecular interactions which are repulsive in nature and present electrostatic barriers to bond rotation. They are apparently highly ordered polymers, although x-ray diffraction patterns have not satisfactorily indicated crystallinity. The question of crystallinity and order is affected by polymer tacticity which is even more complicated by the comonomers employed. The literature shows polyacrylonitriles (PAN) to range from an almost completely atactic material (6–8), to an isotactic structure with crystallinity(9), to a polymer 80% syndiotactic by x-ray measurements(10). One study using NMR spectroscopy shows that PAN contains isotactic and syndiotactic units in a ratio of 50:50(11). This conclusion was reached after studying three samples which were prepared at different temperatures in the range of −78 to 120°C. Prior to this work, Talamini and Vidotto(12), as well as Chiang(13), observed that PAN samples prepared under a variety of conditions showed differences in dissolution and crystallization behavior. These differences were attributed to probable variations in the degree of stereoregularity.

C. Fiber Formation

The formation of a fiber from these polymers can be accomplished by either a dry- or wet-spinning process. The polyacrylics can be prepared by suspension, emulsion, or solution polymerization methods; however, there is some advantage in the solution method where the polymerization solvent can be used satisfactorily as the spinning solvent.

In the wet- and dry-spinning methods the solvents range from highly polar organic compounds to concentrated aqueous solutions of inorganic salts. The organic solvents include dimethylformamide, dimethylacetamide, dimethylsulfoxide, succinonitrile, ethylene thiocyanate, and others. The salts include sodium and calcium thiocyanates and zinc chloride as well as inorganic acids. A comparison of the solvents indicates that dimethylformamide is most efficient and that this, as well as dimethylsulfoxide, produce a polymer solution whose viscosity is stable for 24 h at 100°C (14). The supramolecular structure of the fiber is affected by the method used for solvent removal. In dry spinning there is the removal of solvent by heat and in wet spinning there is the diffusion of the solvent or aqueous salt solution from the coagulated fiber. If the polymer is dissolved in a concentrated aqueous salt solution, it is spun into a more dilute aqueous solution of the same salt for coagulation. If the polymer solvent is organic, the wet-spun fiber may be formed by extrusion into an aqueous solution of the polymer solvent. Temperature is very important in the wet-spinning process in determining the quality of the fiber, and low temperatures, 10°C and below, seem to be necessary for an optimal balance of fiber properties. Wet-spun fibers usually have either a round or lima-bean cross section, while dry-spun fibers are generally dog-bone shaped or flat. Obviously, considerable differences in performance properties as well as aesthetics can be attributed to these differences in cross section.

Knudsen(15) has studied the influence of coagulation variables on the structure and resulting physical properties of acrylic fibers produced by a wet-spinning process. Within this research, a general conclusion is that slow coagulation is better than rapid coagulation. Slow coagulation can be controlled by lowering both temperature and bath composition. The physical properties respond to these variables as well as solids content of spinning solution. The effect of comonomers, spinning, and drawing conditions and processing temperatures on these and other properties of polyacrylonitrile fibers have also been examined by a number of other workers(*16–20*).

There appears to be no (real?) disagreements on the structural concept that polyacrylonitrile is laterally ordered as a single phase combining the properties of both crystalline and amorphous polymer behavior. This is illustrated in studies on molecular orientation in acrylic fibers and changes in the orientation upon stretching(*21*).

D. Fiber Properties

1. *Chemical*

In general, the chemical stability of acrylic fibers are good. They show only poor resistance to strong alkalies, very strong acids, and to a few polar organic solvents. Thus, good resistance is shown toward strong and weak acids and weak alkalies. They are unaffected by common organic solvents. The fibers also show excellent resistance to sunlight and biological attack.

2. *Physical*

The physical properties of acrylic fibers vary within the generic family. As stated earlier, polymers containing one or even two other monomers in small quantities were designed for specific purposes and variation is expected. The general property ranges are shown in Table I.

TABLE I

Property	Property range
Tenacity, g/den.	
conditioned	2.0–3.6
wet	1.6–3.0
Elongation, %	
conditioned	20–50
wet	25–60
Initial modulus, g/den.	40–50
Elastic recovery, %	90–95 at 1% extension
Moisture regain, %	1–3
Specific gravity	1.16–1.18
Thermal property	sticks at 420–490°F

In addition, polyacrylonitrile displays a glass transition between 85 and 100°C which appears to vary with molecular weight. It also seems to have at least a second transition above 100°C(22) but the reported values show considerable variation.

III.   Chemical Reactions

A. EFFECT OF HEAT, OXYGEN, ACID, AND ALKALI

Color can be produced during the polymerization of acrylonitrile as well as in the processes of spinning, washing, drying, and further finishing of the fiber. The chromophore responsible for this behavior in acrylics has been investigated by Burlant and Parsons(23), and Grassie and McNeill(24), among others(25, 26). Miyamichi and coworkers (26), for example, studied the changes in molecular structure in the temperature range of 200–1000°C. Their observations include the pyrolysis and carbonization processes producing a carbon fiber. The intermediates formed at pyrolysis temperatures are those suggested by previous work(23–25), which are the following:

Intermediate formed in an oxidative atmosphere

Intermediate formed in an inert atmosphere

A number of crosslinking reactions apparently occur in the formation of these six-membered rings. And it is from these six-membered rings that graphite nuclei(27) arise in the production of carbon fibers from acrylonitrile polymers. The chemical reactions involved in color formation are complex and Fig. 1 is a typical DTA curve(28) showing the fiber's behavior over a broad temperature range. The latest and most thorough investigation of color formation which extends the earlier suggested mechanisms, has been carried out by Peebles, Brandrup, Kirby, and Friedlander. They have published a series of papers discussing the origin of the defects causing color and presenting good evidence for them(29–34).

In these studies thermal degradation structures produced below 150°C are considered in order to avoid the nonflammable structure which results at higher temperatures in the carbonizing process. Their

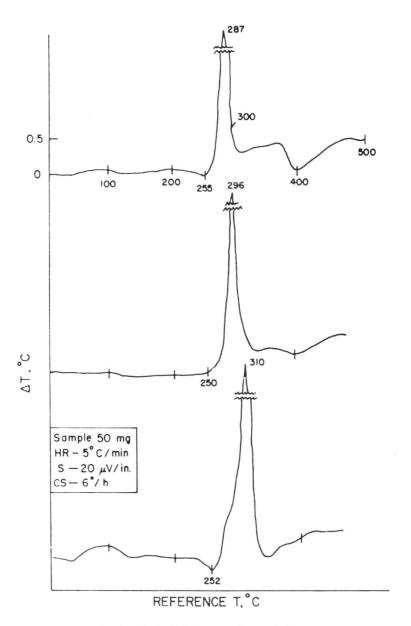

Fɪɢ. 1.   Typical DTA curves for acrylic fibers.

model experiments show that the tertiary hydrogen atoms in nitrile-containing compounds are stable and have no initiating properties when heated under nitrogen. However, in the presence of oxygen, a peroxide and ketone formation initiates color development. The color initiating species may be introduced into the molecule according to the following outline.

(1) *Path A*, major, a β-ketonitrile defect in the chain introduced during polymerization.
(2) *Path B*, major, an attack by oxygen.
(3) *Path C*, minor, an attack by oxygen on the tertiary hydrogen to form a hydroperoxide.
(4) *Path D*, minor, hydrolysis of the nitrile group by acids and bases to form amide and carboxy groups.
(5) *Path E*, major, a nucleophilic attack on the nitrile group carried out by the nucleophilic agents produced by the above (A–D) paths which is responsible for formation of the chromophore.

In Path A, the β-ketonitrile (**I**) is formed by addition of a growing polymer chain end to a nitrile group of a polymer molecule forming an enamine(**II**) which then hydrolyzes.

Path A

The cyanoenamine (**II**) is stable in organic solvents and has a strong uv absorption at 265 μ while the β ketonitrile (**I**) is a titratable weak acid with its ionized form absorbing strongly at 275 μ. The 275 μ absorption can be used to measure the ketonitrile content of polymers photometrically.

Path B is the proposed major route of thermal oxidative degeneration in which attack is at the β-hydrogen atom instead of the tertiary hydrogen. This is preferred because of the polar ·CN group and thus differs from the oxidation in hydrocarbon polymers. A hydroperoxide is formed in this route and breaks down with the elimination of water to form a β-ketodinitrile.

$$
\begin{array}{ccc}
-\text{CHCH}_2-\text{CH}- & \xrightarrow{\text{O}_2} & \overset{\text{OOH}}{\underset{}{-\text{CHCH}-\text{CH}-}} \\
\quad\ |\quad\quad\ | & & \quad\ |\quad\quad\ | \\
\quad\text{CN}\quad\ \text{CN} & & \quad\text{CN}\quad\ \text{CN}
\end{array}
$$

$$
\xrightarrow{\text{H}_2\text{O}}
\begin{array}{c}
\quad\quad\ \text{O} \\
\quad\quad\ \| \\
-\text{CHC}-\text{CH}- \\
\ |\quad\quad\ | \\
\text{CN}\quad\ \text{CN}
\end{array}
$$

Path B

Hydroperoxide formation via oxygen attack at the tertiary hydrogen is Path C, and Path D is the hydrolysis of the nitrile group by acids and bases. These reactions are considered to be minor. Takata and co-workers(35) have also demonstrated color formation through the formation of amide and carboxyl groups.

All the nucleophiles which are produced by the above routes can attack the nitrile group to form a chromophore via Path E. The major nucleophile producers, Paths A and B, produce ketonitriles and these will attack the next nitrile group to form a six-membered ring. Thus, the formation of color is attributed to this partially hydrogenated naphthyridine-type ring produced by the linking up of adjacent nitrile units. The resulting structure can absorb some oxygen to form poly-nitrone $(-\text{C}{=}\text{N}({\rightarrow}\text{O})-)-_n$ units. In this manner Peebles *et al.* explain the mechanism of color formation in acrylonitrile polymers and the major role of heat and oxygen in its degradation.

In order to prevent color formation a number of "stabilizers" for both thermal and photochemical discoloration have been effective. These include such compounds as potassium titanium oxalate(36), N-substituted maleamic acid or its alkali metal salt(37), ethylenediamine tetraacetic acid(38), halogenated phosphines(39), thiourea dioxide(40), sulfites (41), alkyl boronates(42), and naphthalenedisulfonic acids(43). From the color-forming mechanisms which have been outlined, one can postulate the role that some of these stabilizers play in preventing formation of the chromophore.

B. CROSSLINKING AND GRAFTING

Wrinkle recovery, set, shrinkage, and crimp are aspects of acrylic fibers that are usually controlled by the choice of comonomer and the

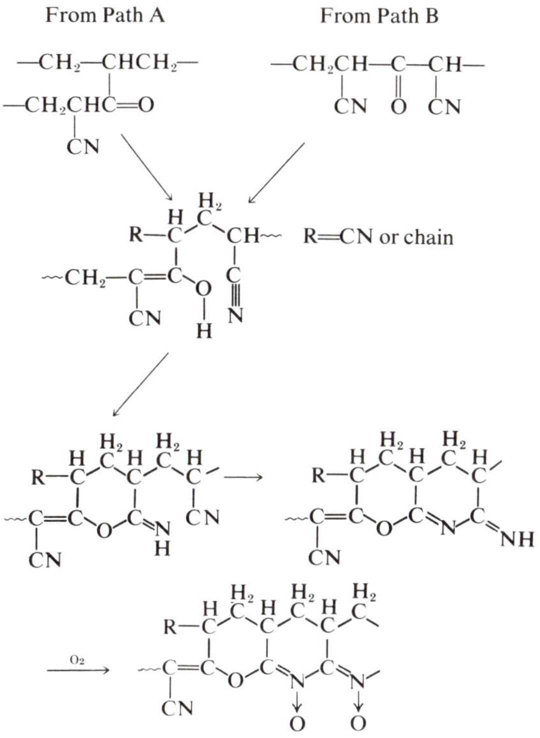

From Path A

$-CH_2-CHCH_2-$

$-CH_2CHC=O$
$\qquad\quad |$
$\qquad\quad CN$

From Path B

$-CH_2CH-C-CH-$
$\qquad\ |\quad \|\quad |$
$\qquad\ CN\ \ O\ \ CN$

$$R-C\underset{|}{\overset{H}{}}\overset{H_2}{\underset{\ }{C}}CH\sim \qquad R=CN\ or\ chain$$

$\sim CH_2-C=C$
$\qquad\quad |$
$\qquad\quad CN$

Path E

$$\xrightarrow{O_2}$$

processing of the fiber immediately after the fiber-forming operations or after fabrication. The wrinkle recovery of a fabric can be improved by heating under constant length and width at 200–220°C for a short time(*44*), or setting can be accomplished by the application of a swelling agent before or after the desired deformation. The use of alkylamides(*45*) and cyanoalkylamides(*46*) are reported to be useful in achieving this type of fiber deformation.

The modification or control of fiber properties in acrylic fibers by crosslinking has been to date principally a chemical exercise. Crosslinking can be accomplished by irradiation(*47–49*) or by incorporation of a reactive comonomer. Among the comonomers that have been reported recently which furnish potential crosslinking sites are (1) methacrolein(*50*), (2) N-methylol acrylamide(*51*), (3) glycidyl methacrylate(*52*), (4) diketene(*50*), and (5) methylenebisacrylamide(*53*).

The fiber property changes, which can be obtained by the introduction of crosslinks, depend on the state of the fiber at the time the crosslinks are formed. When a partially crosslinked polymer is spun into a fiber(*53*) improved tear strength, dyeability, and good elasticity are claimed. When the crosslink is formed after crimping, the crimp is fixed. Strength and elongation are usually decreased somewhat, but thermal stability is frequently increased considerably. Polyacrylic knitted or woven fabrics may be given aftertreatments with methylolating polymers for improved crease resistance.

The acrylic polymers can be grafted before spinning the fiber or after the fiber is formed. If grafting is carried out before spinning, the polymer will have modified solubility properties as well as physical properties which are different from a grafted spun fiber. A summary of this work is given by Battaerd and Tregear(*54*) and covers the early work on the homopolymer poly(acrylonitrile) to improve its antistatic and dyeability properties.

C. OTHER MODIFICATIONS

1. *Flameproofing and Heat Stability*

Flammability can be controlled somewhat by the comonomers, and these monomers contain the familiar elements associated with flameproofing of other fibers and fabrics, particularly the cellulosics. Improvement in thermal stability is desired because the polyacrylics decompose before melting. During the decomposition yellowing is experienced as previously discussed in some detail.

A compound such as vinyl bromide(*55*) may be used as a comonomer for improved flame retardancy. Alternatively, chemicals such as an aromatic chloro or bromo compound with an organic phosphorous compound, both water insoluble, can be dispersed in the polymer solution prior to spinning(*56*). Other flame resistance formulations employ a polybromocyclohexane(*57*) or a brominated phosphonate and a water-insoluble calcium phosphate(*58*) suspended in the polymer solution. The ratio of calcium phosphate to bromine should be 1 : 5.5– 0.7. There appears to be a synergistic effect between the phosphate and the halo compound. This is reminiscent of the effect of sulfide in the melt stabilization of polypropylene. Other methods for improved thermal stability include extensive treatment of the acrylic fiber or fabric with ammonium polysulfide(*59*) or treatment of an acrylic fiber containing aldehyde groups with dimethoxy phosphonite in the presence of sodium hydroxide or ammonium hydroxide to form phosphonic acid esters(*60*).

Improved heat stability may also be obtained by the inclusion of zirconium salts such as the acetate in the spinning solution(*61*).

$$
\begin{array}{c}
\text{H} \\
| \\
-\text{CH}_2\text{C}- \quad\quad \text{O} \\
| \quad\quad\quad \nearrow \\
\text{HC}-\text{P}-(\text{OCH}_3)_2 \quad\quad \text{Phosphonic acid ester} \\
| \\
\text{OH}
\end{array}
$$

2. *Antistats*

As do many synthetic fibers, the acrylic fibers develop static electrical charges which are not dissipated as readily as desired. This may be attributed to their low moisture content due to the lack of groups capable of hydrogen bonding water molecules. Improved antistatic properties are usually obtained by aftertreatments on the fiber or fabric. These formulations consist of chemicals contributing the following physical and chemical properties:

Wetting
    Example: a polyglycol ether
Lubrication
    Example: fatty acids, polysiloxanes
Hydroxyl and/or amino groups
    Example: glycerol, polyhydric alcohols, diethylene triamine,
    or polyamine, or a deliquescent salt.
Water-soluble or dispersible polymeric anchoring agent
    Example: polyamine, -amide, or epoxy resin
A source of —CHO, or if epoxy resin the appropriate crosslinking
agent with catalyst.

Variations in the formulation of antistats are numerous and are virtually trade secrets, although recent patents(62–69) in this field, which exemplify the above characteristics, are available. Static improvement is obtained by stretching freshly spun fibers in hot (95–135°C) ethylene glycol(70), or by prolonged treatment in hot (130–160°C) acidified ethylene glycol for 7 h(71). Such treatments allow good diffusion of the glycol for the desired humectant effect but are not durable to laundering or dry cleaning. The wet-spinning process for acrylics produces a swollen, porous fiber that can absorb a hydrophilic plasticizer, as well as other chemicals including chlorophenols (72), and hydroxy aromatic compounds as bacteriostats.

D. DYEABILITY

There has been a joint effort by the dyestuff manufacturer, the fiber producer, and the dyer to finisher to develop the variety of colors with good fastness and brilliance of shade available in acrylic fabrics today.

The fiber producer has contributed to this success by modification of the basic polymer. This improvement is brought about by the incorporation or creation of acidic groups in the polymer and use of cationic or basic dyestuffs. The acidic group frequently used is the sulfonic acid or its salt which is carried into the polymer chain as a substituent of a vinyl benzene, allyloxy benzene, or diamino stilbene monomer(*73–76*). Also, while the fiber is in the gel state, it can be treated with a sulfonic acid derivative(*77*) and upon drying and scouring it will retain sufficient acid groups for cationic dyeability. If the third monomer is a vinylpyridine(*78–79*) or methylvinylpyridine, or contains a tertiary or quarternary nitrogen(*80*) in the molecule, substantivity to acid dyes may result. A polar, nonionizable monomer containing functional groups, such as an alcohol or ketone, affords sites for complexing with certain dyes. Dyeability with dispersed dyes is also improved by almost any comonomer which alters the structure and reduces the density or compactness of the fiber structure.

IV. Modacrylics

Since these fibers may contain 35–85% acrylonitrile by weight, there can be a wide variation in quantities of comonomers and a resulting variation in properties, particularly dyeability. Rutley has given a general outline of these fibers and their compositions(*81*). Comonomers include vinyl chloride, vinylidine chloride, vinyl acetate, vinyl pyridine, vinyl pyrrolidone, and methyl acrylate or methyl methacrylate. They may be spun by a wet- or dry-spinning method and in dry spinning cheaper and lower boiling solvents, such as acetone, may be used. The better known modacrylic fibers have a ribbon-shaped or a peanut-shaped cross section. Their general physical properties are outlined below in Table II.

TABLE II

Property	Range
Tenacity, g/den.	2.0–4.2
Elongation, %	17–42
Specific gravity	1.28–1.37
Thermal properties	Shrinks at 260–300°F

Both dry and wet strengths are similar with good dimensional stability, resilience, rapid drying, and warmth properties. They usually display a high resistance to combustion, chemical degradation, moths, and mildew. They are stable to uv light but are rather sensitive to heat. They may shrink, soften, or lose tenacity and resilience at temperatures of 120°C and up. One problem encountered with the modacrylics is the loss of luster at the boil; thus, dyeing must be carried out at lower temperatures and frequently the fiber must be relustered. This may be achieved by (1) dry heat, (2) steam heat, or (3) treatment in hot salt solutions(*81, 82*).

The loss of luster has been attributed to the lattice nature of the fiber. This lattice effect results from the large proportion of comonomer in the polymer backbone causing it to behave as essentially two different polymers. This condition promotes light scattering within the lattice structure assumed by the fiber. These fibers have a somewhat reduced affinity for cationic dyes because of their lower acrylonitrile content but an increased affinity for disperse dyes. If the polymer contains a nitrogen derivative such as vinyl pyridine, it will show good dye-ability with acid, chrome, and metal-complex dyes.

A few examples of modacrylic-type polymers, which may be referred to in the literature as interpolymers, are composed of acrylonitrile, vinyl bromide, and vinyl or vinylidene chloride(*83*) or acrylonitrile, vinylidene chloride, vinyl acetate, and the sodium salt of p-methally-oxybenzenesulfonate(*84*) or a substitution of methyl acrylate or styrene for the vinyl acetate(*85*).

REFERENCES

1. *Textile Orgánon*, **XLI** (2), 33 (1970).
2. J. J. Press; ed., "Manmade Textile Encyclopedia," Wiley-Interscience, New York, 1959, p. 35.
3. C. W. Davis and Paul Shapiro, *Encycl. Polym. Sci., Technol.*, **1**, 342 (1964).
4. D. R. Paul, *J. Appl. Poly. Sci.*, **12**, 2273 (1968).
5. R. A. Polatovskaya, E. A. Pakshver, and A. B. Pakshver, *Karbosepnye Volokna*, **1966**, 166.
6. C. L. Arcus and A. Bose, *Chem. Ind. (London)*, 1956, 456.
7. C. Y. Liang and S. Krimm, *J. Poly. Sci.*, **31**, 513 (1958).
8. C. R. Bohn, J. R. Schaefgen, and W. C. Statton, *J. Poly. Sci.*, **55**, 531 (1961).
9. Yahide Kotake, Toshio Yoshihara, Hiroshi Sato, Nobuo Yamada, and Yasushi Joh, *J. Poly. Sci., Part B*, **5**, 163 (1967).
10. R. Stefani, M. Chevreton, and C. Eyraud, *Compt. Rend.*, **251**, 2174 (1960).
11. G. Svegliado, G. Talamini, and G. Vidotto, *J. Poly. Sci., Part A-1*, **5**, 2875 (1967).
12. G. Talamini and G. Vidotto, *Chim. Ind. (Milan)*, **46**, 371 (1964).
13. R. Chiang, *J. Poly. Sci., Pt. A*, **3**, 2019 (1965).

14. V. L. Tsiperman, A. B. Pakshver, and E. A. Pakshver, *Karbotsepnye Volokna,* **1966,** 158.

15. J. P. Knudsen, *Text. Res. J.,* **33,** 13 (1963).

16. M. Larticle, S. G. Lefebvre, and F. A. Eeckman, *Ann. Sci. Textiles Belges* (2), 7 (1968).

17. Kazuhisa Saito, *Sen-i Gakkaishi,* **24,** 323 (1968).

18. Masao Takahashi, Yasuhiko Nubushina, and Setsuko Kosugi, *Text. Res. J.,* **34,** 87 (1964).

19. Hiromu Takeda and Yasuhiko Nukushina, *Text. Res. J.,* **34,** 173 (1964).

20. Shlomo Rosenbaum, *J. Appl. Poly. Sci.,* **9,** 2071 (1965); ibid., 2085 (1965).

21. Nobuhiro Tsutsui, Takuro Hayahara, Yosuo Matsumura, and Hideto Sekiguchi, *Kobunshi Kagaku,* **23,** 193 (1965); ibid., 199.

22. Saburo Okajima, Morio Ikeda, and Akio Takeuchi, *J. Poly. Sci., Pt. A-1,* **6,** 1925 (1968).

23. W. J. Burlant and J. L. Parsons, *J. Poly. Sci.,* **22,** 249 (1956).

24. N. Grassie and I. C. McNeill, *J. Poly. Sci.,* **27,** 207 (1958); ibid., **39,** 211 (1959).

25. N. Grassie and J. N. Hay, *J. Poly. Sci.,* **56,** 189 (1962).

26. Kazuo Miyamichi, Masao Okamoto, Osamu Ishizuka, and Masamichi Katayoma, *Sen-i Gakkaishi,* **22,** 538, 548 (1966).

27. W. Watt and W. Johnson, *Appld. Poly. Symp.,* No. 9, 215 (1969).

28. Robert Schwenker, Personal Products, Div. of Johnson and Johnson, Milltown, N.J.

29. L. H. Peebles, Jr., and J. Brandrup, *Makromol. Chem.,* **98,** 189, (1966).

30. L. H. Peebles, Jr., *J. Poly. Sci., A-1,* **5,** 2637 (1967).

31. J. R. Kirby, J. Brandrup, and L. H. Peebles, Jr., *Macromol.,* **1,** 53 (1968); ibid., **1,** 59 (1968).

32. J. Brandrup and L. H. Peebles, Jr., *Macromol.,* **1,** 64 (1968).

33. J. Brandrup, *Macromol.,* **1,** 72 (1968).

34. H. N. Friedlander, L. H. Peebles, Jr., J. Brandrup, and J. R. Kirby, *Macromol.,* **1,** 79 (1968).

35. Toshihiro Takata, Iwao Hiroi, and Masakazu Taniyama, *J. Poly. Sci., Pt. A2,* 1567 (1964).

36. Netherlands Application 6,600,566, Courtaulds, Ltd.

37. Netherlands Application 6,605,901, Courtaulds, Ltd.

38. U.S. Patent 3,415,611, Dow Chemical.

39. Japan Patent 01,093(68), Toho Rayon.

40. French Patent 1,439,365, SNIA VISCOSA.

41. Japan Patent 15,312(66), Toho Rayon.

42. Japan Patent 9,629(65), Toyo Rayon.

43. French Patent 1,374,840, Monsanto Chemical Co.

44. U.S. Patent 3,395,037, Monsanto Company.

45. British Patent 929,399, Courtaulds.

46. British Patent 954,157, Courtaulds.

47. W. J. Burlant and C. R. Taylor, *J. Phys. Chem.,* **62,** 247 (1958).

48. L. Bateman, M. Cain, T. Colclough, and J. I. Cunneen, *J. Chem. Soc.,* 3570 (1962).

49. Icksam Noh, *Daehan Hwahak Hwoeju,* **11** (2), 77 (1967) [C.A., **68,** 10212 (1968)].

50. A. A. Konkin, *Przemysl Chem.,* **45** (5), 233 (1966).

51. British Patent 1,251,902, Courtaulds, Ltd.

52. S. Kamalov, A. F. Gladkikh, N. N. Ivanov, G. A. Shtraikhman, and S. Ya Frenkel, *Khim. Volokna,* (3) 21 (1967).

53. German Patent 1,272,485, Fosfatbolaget AB.

54. H. A. J. Battaerd and G. W. Tregear, "Grafted Copolymers," Wiley-Interscience, 1967, p. 257.
55. Netherlands Application 6,517,131, Monsanto.
56. French Patent 1,523,199, American Cyanamid Co.
57. U.S. Patent 3,213,052, Dow Chemical.
58. U.S. Patent 3,242,124, Dow Chemical.
59. Netherlands Application 6,514,477, Farbenfabriken Bayer.
60. Z. A. Rogovin, M. A. Tyuganova, G. A. Gabrielyan, and N. F. Konnova, *Khom. Volokna*, (3) 27 (1966).
61. U.S. Patent 3,296,171; 3,296,348, Dow Chemical Co.
62. Belgian Patent 664,438, Farbenfabriken Bayer.
63. Belgian Patent 664,255, Farbenfabriken Bayer.
64. Belgian Patent 665,163, Farbenfabriken Bayer.
65. Netherlands Application 6,518,684, Monsanto Co.
66. French Patent 1,442,735, American Cyanamid Co.
67. German Patent 1,211,577, Farbenfabriken Bayer.
68. U.S. Patent 3,376,245, Monsanto Co.
69. U.S. Patent 3,423,314 Dow Corning Corp.
70. L. A. Yasnikov and E. A. Pakshver, *Khim. Volokna*, (4), 54 (1968).
71. German Patent 1,276,589, Textilforschungsanstalt Krefeld.
72. Netherlands Application 6,507,931, Dow Chemical; U.S. Patent 3,312,758, Dow Chemical.
73. U.S. Patent 3,380,798, Monsanto Chemical.
74. U.S. Patent 3,402,014, Monsanto Co.
75. Z. I. Burlyuk, N. M. Beder, and B. K. Kruptsov, *Karbotsepnye Volokna*, **1966**, 44.
76. French Patent 1,443,031, Eastman Kodak Co.
77. U.S. Patent 3,402,014, Monsanto Co.
78. U.S. Patent 2,491,471; 2,916,348, DuPont.
79. British Patent 866,982, American Cyanamid.
80. British Patent 865,814, Courtaulds, Ltd.
81. R. O. Rutley, *Text. Mfgr.*, **94**, 211 (1968).
82. H. Kellet, *J. Soc. Dyers Colour.*, **84**, 257 (1968).
83. Netherlands Application 6,517,132, Monsanto Co.
84. Netherlands Application 6,517,190, Monsanto Co.
85. Netherlands Application 6,517,189, Monsanto Co.

Jean/Jean Top/Shirt/Overshirt
Sorter

Bankcard
Knickers
Tampon
Money
T-Shirt
Nightie
Towels

Wash Hair
Shoes
Jeans Top
Pencil cases

Card

Chapter 7 POLYETHYLENE TEREPHTHALATE

I. Introduction

Polyethylene terephthalate (PET) is a linear polyester consisting in the truest organic sense of repeating ester groups $-C(=O)-O-$. The usefulness of polyesters for fibers was first revealed by Whinfield and Dickson(1) when they incorporated terephthalic acid into the repeat unit of the polymer molecule. The rapid acceptance of this polymer in fiber form is evidenced by the current annual world production of over a billion lb. Polyethylene tetephthalate and its modification through the introduction of comonomers provides a fiber with a variety of useful properties. It is particularly desirable in fabrics where it has been blended with a cellulosic fiber such as cotton or rayon.

II. Chemical and Physical Structure

A. CHEMICAL STRUCTURE

Polyethylene terephthalate, a good fiber-forming polymer, was first made commercially by the transesterification reaction between dimethyl terephthalate (DMT) and ethylene glycol (EG) followed by a polycondensation reaction as follows:

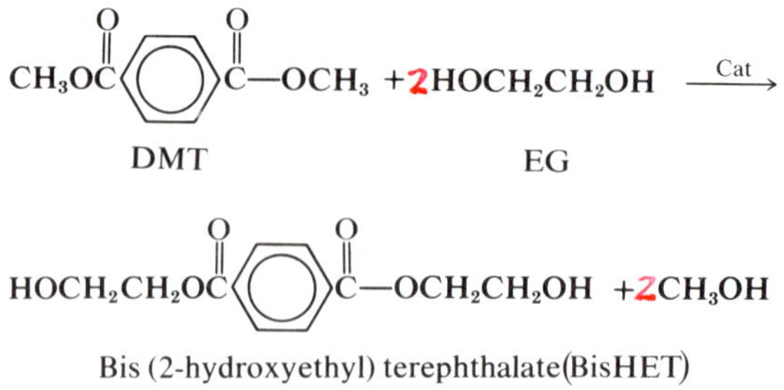

$$CH_3OC\underset{O}{\overset{O}{\|}}\!\!\!\bigcirc\!\!\!\overset{O}{\overset{\|}{C}}\!\!-OCH_3 + 2HOCH_2CH_2OH \xrightarrow{\text{Cat}}$$

DMT   EG

$$HOCH_2CH_2OC\underset{O}{\overset{O}{\|}}\!\!\!\bigcirc\!\!\!\overset{O}{\overset{\|}{C}}\!\!-OCH_2CH_2OH + 2CH_3OH$$

Bis (2-hydroxyethyl) terephthalate(BisHET)

transesterification

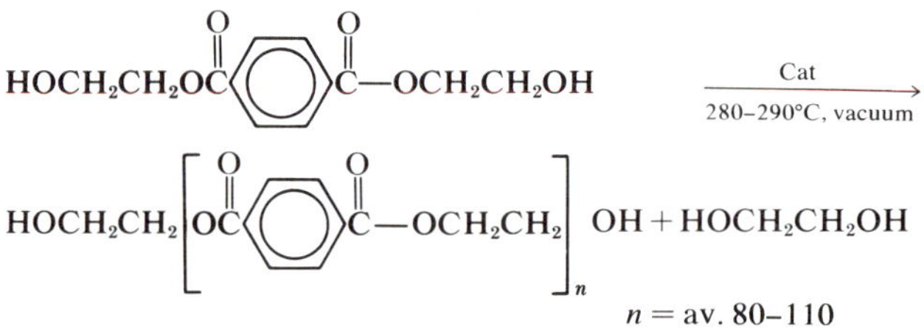

$$HOCH_2CH_2OC\underset{O}{\overset{O}{\|}}\!\!\!\bigcirc\!\!\!\overset{O}{\overset{\|}{C}}\!\!-OCH_2CH_2OH \xrightarrow[\text{280--290°C, vacuum}]{\text{Cat}}$$

$$HOCH_2CH_2\left[OC\underset{O}{\overset{O}{\|}}\!\!\!\bigcirc\!\!\!\overset{O}{\overset{\|}{C}}\!\!-OCH_2CH_2\right]_n OH + HOCH_2CH_2OH$$

$n = $ av. 80–110

polycondensation

The catalyst systems which are used in the dimethyl terephthalate process usually consist of zinc, manganese, calcium, and lead salts for transesterification and an antimony compound for the polycondensation. More recently pure terephthalic acid has become available in commercial quantities. If the free acid is used as the monomer no transesterification catalyst is necessary but the direct esterification with ethylene glycol requires additional heat with pressure. This is necessary to maintain a feasible reaction rate and at the same time to keep ethylene glycol (bp 196°C) from boiling away. In the direct esterification of terephthalic acid, water is the byproduct and there is no methanol disposal problem. In both processes, the ethylene glycol may undergo a dehydration reaction to form diethylene glycol.

$$2\,HOCH_2CH_2OH \longrightarrow HOCH_2CH_2OCH_2CH_2OH + H_2O$$

This compound is found in the backbone of the polymer sometimes to the extent of 2–2.5 mole %, and may be regarded as a comonomer(2). With present-day technology, polyester prepared from dimethyl terephthalate contains less diethylene glycol than that prepared from terephthalic acid. By both processes, however, 1.5–1.7% cyclic oligomers are produced, with the trimer being the principal contaminant(3). The incorporation of diethylene glycol can occur as a mixed cyclic dimeric ester. Polyethylene terephthalate was first produced in a batch process, but continuous polymerization processes using dimethyl terephthalate or terephthalic acid have been developed which promise technical and economic advantages.

Titanium dioxide (anatase) is normally used to deluster the polyester and is usually added at the beginning of the reaction. The amounts used vary from 0.05% for a very small dulling effect to 2% for a full-dull effect. In addition to a well-dispersed TiO_2, triaryl phosphites or phosphates are added to the reaction melt to give melt stability and improved color to the resin. Monofunctional chain terminators can be used to control molecular weight. Fluorescent optical brighteners may also be added during the polycondensation to give added "brightness or whiteness" to the fiber when exposed to sunlight.

B. PHYSICAL PROPERTIES

When quenched rapidly from the melt the polymer is an amorphous, glassy solid. Differential thermal analysis (DTA) shows a second-order transition at 78–80°C, a crystallization endotherm ranging between 125 and 180°C, and a melting point (DTA) of 255°C. Birefringent melting points are in the neighborhood of 260–265°C. The melting point is lowered by an increased diethylene glycol content. Danbeny, Bunn, and Brown(4) have determined the crystal structure from x-ray diffraction studies.

The polymer is insoluble in most common solvents and solution viscosities for molecular weight determinations are made in such powerful solvents as orthochlorophenol, phenol-tetrachloroethane, or dichloroacetic acid. The melt viscosity of polyethylene terephthalate is very high and will reach about half a million cP at a molecular weight of 23,000. Polyester fibers usually have a molecular weight of 18,000–25,000, although for higher strengths, a higher molecular weight might be necessary. Molecular weight distribution corresponds well with the predictions of the Flory distribution function(5). If PET in the solid state is subjected to heat at a reduced pressure, further polycondensation occurs, resulting in a much broader molecular weight distribution (6).

III. Fiber Properties

A. PHYSICAL PROPERTIES

Polyethylene terephthalate fibers are produced by a melt spinning process which requires very dry resin and temperatures of 290–315°C. The as-spun fiber is amorphous and shows only a small amount of molecular orientation as indicated by a small birefringence. In order to develop useful fiber properties, the as-spun yarn is drawn at temperatures ranging from 180 to 200°C (7). The drawn fiber is highly oriented and is characterized by a well-developed crystalline structure. Van Veld, Morris, and Billica, using a scanning electron microscope, have shown a complex heterogeneous fiber structure including a crust and an internal fibrillar matrix (8). The resulting fiber, depending on molecular weight and spinning and drawing conditions, may have staple tensile properties in the ranges given in Table I.

TABLE I

Property	Range
Tenacity, g/den.	2.5–6.0
Elongation, %	12–50
Elastic recovery, %	90–96 at 2% extension
Average stiffness, g/den.	8–25
Average toughness, g cm/den.-cm	0.4–1.5
Specific gravity	1.38
Moisture regain,%	
65% r.h. at 70°F	0.4

Fiber tenacity and the tensile modulus are reduced only slightly when wet. The fiber maintains its strength during prolonged periods of heating. Fiber shrinkage will occur and is linearly related to temperature. A highly drawn yarn will shrink 1–4% in water at 100°C. When the yarn is held at a fixed length during a heat treatment, the load-extension properties are not affected very much, even though such setting or thermal annealing reduces shrinkage from 12–15% to 2% or less. The thermal shrinkage and annealing properties are of prime importance in the stabilization of polyester fabrics (9).

B. CHEMICAL PROPERTIES

The resistance of PET fibers to weak acid, alkali, and normal bleaching agents is good, and strong acid, such as 30% hydrochloric, has little or no effect on fiber strength at room temperatures. However, basic substances, such as caustic soda, will attack the surface of the fiber causing saponification and the production of soluble products, leaving an etched surface in some instances. Organic bases and ammonia will be absorbed into the fiber and the general aminolysis will cause a loss in strength(*10*). The attack of methylamine reveals the fiber structure by penetrating the amorphous or less ordered regions with no penetration of the crystalline regions.

Other than this degradation, which is hydrolytic in essence, PET fibers may degrade by thermal, oxidative, and radiation-induced reactions. Buxbaum(*11*) has considered all four degradative reactions, frequently using model compounds to interpret the reaction kinetics.

1. *Thermal Degradation*

Polyethylene terephthalate subjected to heat between 200–300°C, and particularly above its melting point, degrades by a molecular mechanism involving random chain scission at the ester links. A radical mechanism has been proposed(*12*) but no evidence either in model compounds or bulk polymer has been found to establish this. Pohl(*13*), after studying a variety of terephthalate polyesters, suggested that the principal reaction in the degradation begins at the β-methylene group leading to the following breakdown:

$$\sim C_6H_4\overset{\overset{\displaystyle O}{\|}}{C}-OCH_2CH_2O\overset{\overset{\displaystyle O}{\|}}{C}C_6H_4\sim \xrightarrow{\Delta}$$

$$\sim C_6H_4\overset{\overset{\displaystyle O}{\|}}{C}-OCH=CH_2 + HO\overset{\overset{\displaystyle O}{\|}}{C}-C_6H_4\sim$$

The gaseous products which are produced between 280–306°C are principally CO, CO_2, H_2O, CH_3CHO, and C_2H_4 with traces of methane, benzene, and 2-methyl-dioxolane. Table II gives some degradation rate constants(*14*).

This degradation, however, is most prevalent in the final stages of polycondensation and in the melt spinning process. Organic phosphorus compounds improve the thermal stability presumably by "neutralizing" certain of the metal ions used in catalysis.

2. *Oxidative Degradation*

Buxbaum(*15*) suggests that diethylene glycol chain segments lower the stability of the polyester to oxidation. Presumably this comonomer, formed during the polymerization from ethylene glycol, reduces

TABLE II

Type	Temp., °C	Ester links broken per day per 10^6 g polymer
Thermal	300	1.1×10^{-1}
	282	3.1×10^{-2}
	100	4.0×10^{-8}
Oxidative	100	2×10^{-8}
Hydrolytic	100	5×10^{-5}
	120	5×10^{-4}
Ammolytic	120	1×10^{-2}

crystallinity and accordingly would be located in the accessible non-crystalline regions of the fiber. The initial step is the formation of a hydroperoxide on a glycolic $-CH_2-$ unit as shown in the following scheme:

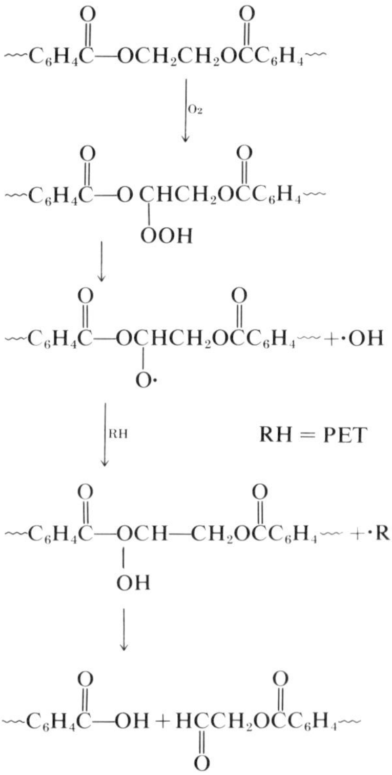

$$\sim C_6H_4\overset{O}{\overset{\|}{C}}-OCH_2CH_2O\overset{O}{\overset{\|}{C}}C_6H_4\sim$$

$$\downarrow O_2$$

$$\sim C_6H_4\overset{O}{\overset{\|}{C}}-O\underset{\underset{OOH}{|}}{C}HCH_2O\overset{O}{\overset{\|}{C}}C_6H_4\sim$$

$$\downarrow$$

$$\sim C_6H_4\overset{O}{\overset{\|}{C}}-O\underset{\underset{O\cdot}{|}}{C}HCH_2O\overset{O}{\overset{\|}{C}}C_6H_4\sim + \cdot OH$$

$$\downarrow RH \qquad\qquad RH = PET$$

$$\sim C_6H_4\overset{O}{\overset{\|}{C}}-O\underset{\underset{OH}{|}}{C}H-CH_2O\overset{O}{\overset{\|}{C}}C_6H_4\sim + \cdot R$$

$$\downarrow$$

$$\sim C_6H_4\overset{O}{\overset{\|}{C}}-OH + H\underset{\underset{O}{\|}}{C}CH_2O\overset{O}{\overset{\|}{C}}C_6H_4\sim$$

$$
\downarrow{\scriptstyle O_2}
$$

$$
\overset{O}{\underset{\|}{HOC}}CH_2-\overset{O}{\underset{\|}{OC}}C_6H_4\sim
$$

or

$$
\sim C_6H_4\overset{O}{\underset{\|}{C}}-O\dot{C}HCH_2\overset{O}{\underset{\|}{OC}}C_6H_4\sim\ +\cdot OOH
$$

$$
\downarrow
$$

$$
\sim C_6H_4\overset{O}{\underset{\|}{C}}-O\cdot + CH_2=CHO\overset{O}{\underset{\|}{CC}}C_6H_4\sim
$$

$$
\downarrow{\scriptstyle RH}
$$

$$
\sim C_6H_4\overset{O}{\underset{\|}{C}}-OH + \cdot R
$$

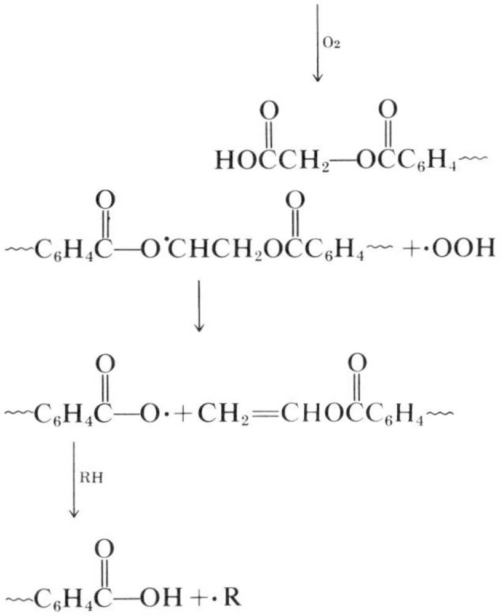

3. Radiation-Induced Degradation

Polyethylene terephthalate is stable to uv light; however, after extensive exposure to wavelengths of 2537 Å and 3000–3300 Å some chemical changes can be observed(*16*). These consist essentially of the formation of free radicals which lead to chain cleavage and crosslinking.

$$
\cdot OCH-CH_2-O- \qquad -O\overset{O}{\underset{\|}{C}}\overset{\dot{}}{\bigcirc}\overset{O}{\underset{\|}{C}}-O-
$$

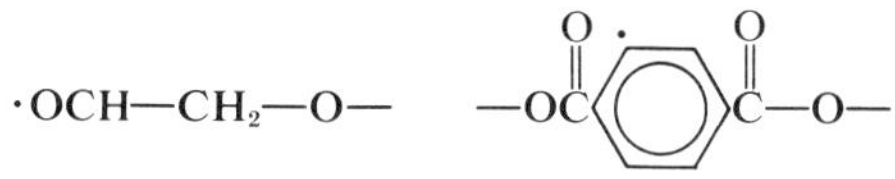

Further study has also identified the monohydroxylated and di-hydroxylated terephthalate moieties(*17*).

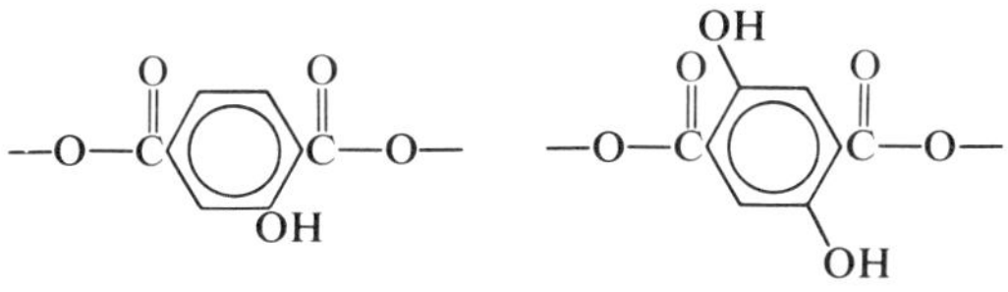

$$
-O-\overset{O}{\underset{\|}{C}}\underset{OH}{\bigcirc}\overset{O}{\underset{\|}{C}}-O- \qquad -O-\overset{O}{\underset{\|}{C}}\overset{OH}{\underset{OH}{\bigcirc}}\overset{O}{\underset{\|}{C}}-O-
$$

Gamma-radiation sources again lead to chain scission via free-radical formation with the predominant radical being $-O-\dot{C}HCH_2-O-$. The mechanical properties change little at moderate doses of high-energy radiation and only show significant strength and elongation losses when the dose exceeds 100 Mrads. Chain cleavage and cross-linking occur depending on the specific reaction conditions. Besides the dose rate, the kind of radiation, temperature, crystallinity, and the presence of impurities such as water or oxygen are influential in controlling the extent of degradation.

IV.   Fiber Property Modifications

The modification of polyethylene terephthalate may be accomplished by the incorporation of small amounts of comonomers into the linear polyester. This may change the fiber's physical and chemical properties as well as necessitate new melt spinning conditions. In this section some of the more important property modifications will be discussed, regardless of whether the chemistry occurs before the formation of the fiber, i.e., in the polycondensation process, to the resin, or in after-treatment of the basic fiber.

A. DYEABILITY

Due to the crystallinity of PET fibers and the hydrophobicity of the polymer, dyeing is difficult to accomplish. Disperse dyes with the aid of a "carrier" are used to obtain the variety and depth of colors that today's market requires. Carriers include compounds such as o-chlorophenol, o-phenylphenol, biphenyl, benzoic acid, salicylic acid and methyl salicylate, and their function appears to be that they plasticize the fiber(18). This results in certain changes in the fiber, especially further crystallization which allows more rapid diffusion of the dispersed dye into the fiber. These chemicals also may be considered to swell or loosen the fiber surface structure, thus allowing better dye penetration. In this method dyeing is carried out at the boil. Other dye methods include a pad–high-temperature (390–410°F) dyeing process often combined with a steam treatment for polyester–cellulose blended fabrics. Further marketing needs now call for (1) a deep-dyeing fiber for tone-on-tone effects with regular PET, (2) a fiber which can be dyed without a carrier, (3) a cationic or basic dye-able PET, (4) an anionic or acid dyeable PET, and (5) a fire-retardant PET.

1. *Comonomers*

Dyeability can be imparted by incorporating a suitable comonomer which will change the character of the fiber's noncrystalline areas.

It has been shown by Hill and Edgar(*19*) that modification of poly-(ethylene terephthalate) by a dibasic acid resulted in adverse effects on mechanical, physical, and chemical properties. In fact, the quantity of a third component that can be tolerated without an unacceptable decrease in melting point or increased fiber shrinkage is small. Also, Flory(*20*) has pointed out that if the length of the third comonomer sequences is sufficiently large, the melting point of the copolymer will be independent of the molar fraction of the third comonomer. This suggests that a poly(ethylene terephthalate) copolymer may be prepared in which the high melting point of PET is retained while a noncrystallizing amorphous area can be introduced that will absorb dispersed dyes.

A review of the literature(*21*) on copolyesters will show that hundreds of copolyesters have been prepared. The comonomers included alkyl diols and dicarboxylic acids, substituted aryl and alkylaryl diols, and dicarboxylic acids as well as a variety of difunctional heterocyclic compounds.

a. *Cationic Dyeability (Basic).* To impart this dyeing property to polyester fibers, comonomers containing salts of sulfonic, sulfenic, and phosphonic and phosphinic acids appear acceptable. The sodium salt of 5-sulfo-isophthalic acid(*22*) at the 2.5-mole-% level gives sufficient ionic dye sites for a commercial fiber with good cationic dyeability. 3,5-Dicarboxylphenyl phosphonic acid or its salt is an example of a difunctional phosphorous compound that is claimed for basic dyeability(*23*).

b. *Anionic Dyeability (Acid).* Polyesters with an affinity for acid dyes, can be prepared by the use of comonomers containing nitrogen. Tertiary amine compounds containing either alkyl, aryl, aralkyl, or cycloalkyl groups have been claimed(*24*). Difunctional triazoles(*25*) such as 1-(carbethoxy-methyl)-4-(hydroxymethyl)-1,2,3-triazole may also be useful in this area.

2. Grafting

Grafting reactions on fibers can bring about a variety of chemical property changes, depending on the monomers used for grafting. Theoretically, the types of dyeabilities discussed earlier in the section can be achieved by grafting(*26*). N-vinyl lactams, vinyl-benzensulfonic acids, p-vinylbenzylamines, and vinylphenyl polyglycol ethers(*27*) are a few of the monomer types which can control the type and extent of dye acceptance in polyester fibers.

3. *Other Treatments*

Surface treatments, particularly on undrawn fiber, may also offer a route for varying dyeability. A liquid polyamide (*28*) at 110–130°C penetrates well into an undrawn fiber, and after drawing enables the fiber to accept acid dyes. Another process, which involves treating with NOCl under irradiation to induce nitrosation (*29*), claims both basic and acid dye acceptability. Surface treatment of a polyester fabric with 1-cyclohexylaziridine (*30*) and a catalyst gives dyeability with acid dyes, or treatment of fibers with dihalogenated alkyl compounds (*31*) improves dispersed dye uptake.

Cospinning of dye-acceptable polymers with polyester is another modification route but one which seemingly has not met with much success.

B. ANTISTATIC TREATMENTS

The accumulation of an electrical charge on fibers and fabrics causes difficulties in processing, possible fire and explosion hazards, and is a nuisance in garments due to their clinging to the body of other garments, and due to electrical shocks through rapid discharging. In view of the hydrophobic character of polyester the absorption of moisture from the atmosphere is so low that a conducting layer on the surface of the fiber does not form. Antistatic agents are chemicals that may either reduce the generation of electrical charges on the fiber or increase the surface conductance of the material to which they are added. Some antistatic agents may function by both pathways. The efficient antistatic agent should be present on the surface of a fiber and should form a continuous conductive coating, which may be regarded as an aqueous solution of a high ionic concentration.

Durability of an antistatic agent to laundering is of importance to the consumer while nondurable agents are adequate for use during spinning and weaving processes. It has been observed that loss of antistatic protection may occur on storage by migration of the agent into the interior of the fiber (*32*). This migration destroys the continuity of the surface coating and a corresponding loss in antistatic protection is observed. Temperature, humidity, volatility, stability (oxidative), chemical structure of the agent, and the compatibility of the agent with the fiber surface determine the storage durability of antistatic agents on polyester fibers and fabrics.

Good aging properties and no migration in polyester–rayon blends are claimed for a formulation containing a lauryl phosphate, morpholine, various polyoxyethylenes, a silicone, and o-phenylphenol [a swelling agent for polyester, frequently used as a dyeing assistant or carrier (*33*)]. Such combinations would produce antistatic properties and also antifrictional properties but would not be durable to laundering. In order to achieve durability in laundering, antistatic finishes for polyesters have recently been developed which are polymeric and frequently contain a few crosslinks. Valko and Tesoro(*34*) have discussed polyamine resins as antistatic finishes. These materials are three-dimensional ion-exchange resins which are hygroscopic and retain the necessary mechanical strength to withstand abrasion during laundering. The crosslinking reaction for three dimensionality is usually carried out after application to the fiber or fabric. The ion-exchange property necessary for conductivity may also be obtained by a cationic poly-ethoxy electrolyte cured with a aziridinyl compound such as tris(1-aziridinyl) phosphine oxide, or a cationic thermosetting polyamide-epichlorohydrin resin(*35*).

These materials are surface films on fibers and their antistatic properties are independent of the type of fiber and the crosslinking method. Seidel(*36*) has studied a number of crosslinking systems with poly-amines containing primary and secondary amino groups, and poly-ethylene oxide residues. For the best permanent antistatic effect, the amine components are not the deciding factor, rather the structure of the chlorohydrin crosslinking component becomes important. Other antistats for polyester include polysiloxanes(*37*) which are then cross-linked with benzyl peroxide(*38*) on the fabric. Some antistats, such as copper, zinc, and nickel salts of alkyl or poly(oxyethylene) phosphates, are incorporated during the fiber-spinning process(*39*).

The quantity of material required for antistatic protection depends on the action of the antistat under the various conditions of temperature and humidity, on fiber linear density, and fabric structure. Generally, and to guard against the possibility of nonuniform treatment, about 0.1% of an antistat is applied. Polyester fabrics show electrical resis-tivities in the range of $10^{13-14}\Omega$ and the application of an antistat will reduce this to $10^{7-10}\Omega$.

C. ULTRAVIOLET STABILIZATION

The chemistry involved in the degradation of polyesters has been discussed previously. It has recently been shown that weathering in different high-sunlight areas in the U.S., such as Florida and Arizona, may produce different strength losses for the same total radiation

(40). This effect is due to seasonal shifts in the uv intensity and it appears that the location variable is eliminated if total uv radiation rather than total radiation is related to strength loss. For protection against degradation by uv leading to strength losses, polyester fibers may contain compounds, added in the preparation of the polymer or during melt spinning, which prevent or slow down these degradation reactions.

These compounds are basically energy absorbing, are thermally stable, and contain considerable aromaticity. They may be benzotriazoles such as 2-(2-hydroxy-5-β-carboxy-ethylphenyl) benzotriazole *(41)*, 2-benzamidophenyl-2H-benzotriazoles*(42)*, or aryl-1,3,5-triazine derivatives*(43)*. Ultraviolet protection may also be obtained by hindered phenols such as 2,4-dimethyl-6-α-methylcyclohexylphenol *(44)*, o-hydroxybenzophenone with p-oriented hydroxyls or ester-forming groups*(45)*, and cupric salts of a monocarboxylic acid such as cupric stearate*(46)*. Ultraviolet stabilizers may be included in the formulations containing antistats for topical application.

D. Soil Release

Soil release finishes for polyesters must be viewed in the light of how soiling of textiles occurs. Generally, soiling occurs by one or a combination of the following mechanisms: (1) physical contact, (2) static attraction, and (3) redeposition during washing. Fabrics are soiled by transference of soil and body oils (sebum) from the skin, by spills, by atmospheric contamination, or by rubbing against dirty surfaces. Generation of static on fibers leads to increased attraction of fine dirt particles, particularly severe when fiber surface moisture is low.

Redeposition during washing produces a greyness to the polyester fabric and is attributed to removal of soil from the fibers or fabrics which are easily wetted and an adsorption on the fibers or fabrics which show a lesser interfacial tension for the soil. The question of soiling and soil release must be considered not only in 100% synthetic fabrics but also in fiber blends, such as polyester–cellulosic fiber blends. The problem is complicated by various chemical treatments to impart crease resistance and durable-press properties. In obtaining good soil- and stain-release properties for polyester containing fabrics, fiber cross sectional shape and crimp, fabric construction, fabric finish, and type and concentration of detergents to be used are parameters to be considered. In polyester–cotton blends two distinct problems may be identified. One is soil redeposition, and the other is removal of oils from delayed-cure type of permanent-press fabrics. It has been shown that polyesters retain less oils on laundering than do DP-treated cottons (47). It has been suggested that this poor removal of oils or fats from the cotton component is due to its high surface irregularity and that in polyester–cotton blends poor oily stain release is due to the cotton (48). This subject is also discussed in the chapter on cotton.

Both fibers probably contribute to the soil-release problem, and a reduction in the hydrophobicity of the polyester would aid soil release during laundering. This would bring the free energy of the polyester–water interface more in line with the free energy of the cellulose–water interface. Thus, the removal of oily soil from the polyester fiber surfaces by water would be accomplished more easily and efficiently.

The claims in finishing treatments for polyester-containing fabrics often include good antisoiling properties along with good soil removal and stain release properties during laundering. Fluorocarbons appear to be most frequently used for antisoiling treatments, particularly oleophobic stains. Examples of these compounds are perfluoroalkyl-substituted polyamides(49), N-ethyl-N-(β-aminoethyl) perfluoro-octanesulfonamide(50), and grafted perfluoroalcohols(51).

Improved soil resistance as well as soil release and no redeposition appear to be obtained by radiation-grafted acrylic acid polymers and copolymers followed by thermal crosslinking of an aminoplast resin (52) by application of a polyhydroxylated polymer made from terephthalic acid and polyethylene glycol(53).

For blends, carboxymethyl cellulose (CMC) and polyhydroxy polymers may be used, with the cellulosic fiber attaining the electronegative groups of the CMC and the polyester absorbing and/or combining with the polyol(54). Soiling and antistatic properties as well as soil removal are dependent on the surface properties of the fiber. Yet it is possible to disperse, prior to spinning the fiber, polyethylene glycol dibenzoate or polyethylene glycol diacetate in the poly(ethylene terephthalate) (55) and produce a fiber with these properties. Soil-release chemistry is in its infancy, relatively speaking, and its future course depends largely on consumer demand.

E. FLAMEPROOFING

Research in this area for polyester fibers has recently become very active. Significant and pertinent flameproofing chemistry has not yet reached the publication stage but it may be anticipated that the development of this property will be achieved. It is speculative at this time as to which are most likely to be successful. It is possible for example, that flameproofing chemicals will be applied in a cospinning process, in treatment of the undrawn or drawn yarn, or even in the crimping process during staple fiber production. Furthermore, the chemical structure of the polyester may be modified by incorporation of flame-retardant chemicals as comonomers. Flameproofing also may be accomplished by incorporating the retardant during the final finishing of the fabric.

Depending on the test methods, varying quantities of flame retardant chemicals are required. However, Miles and coworkers(56) have stated that in polyester molding resins 12–15% phosphorous is required for fire retardancy. Using a compound which also contained a halogen such as tris-(2,3-dibromopropyl) phosphate would probably reduce this requirement thirty fold. Again, having to refer to molding resins, phosphorous–nitrogen-containing compounds, such as phosphoramidates (57), offer flame resistance in a synergistic manner.

REFERENCES

1. British Patent 578,079.
2. D. R. Gaskill, A. G. Chasar, and C. A. Lucchesi, *Anal. Chem.*, **39**, 106 (1967).
3. I. Goodman and B. F. Nesbitt, *J. Poly. Sci.*, **48**, 423 (1960).
4. R. Danbeny, C. W. Bunn, and C. J. Brown, *Proc. Roy. Soc. (London), Ser. A.*, **226**, 531 (1954).
5. G. J. Howard, "Progress in High Polymers," J. C. Robb and F. W. Peaker, eds., Heywood and Co., London, 1961, Vol. I.
6. C.-Y. Cha, *Polymer Preprints*, **6**, 84 (1965).
7. A. B. Thompson, *J. Poly. Sci.*, **34**, 741 (1959).
8. R. D. Van Veld, G. Morris, and H. R. Billica, *J. Appl. Sci.*, **12**, 2709 (1968).
9. D. N. Marvin, *J. Soc. Dyers Colour.*, **70**, 16 (1954).
10. G. Farrow, D. A. S. Ravens, and I. M. Ward, *Polymer*, **3**, 17 (1962).
11. L. H. Buxbaum, *Angew. Chem., Intl. Ed. Engl.*, **7**, (3), 182 (1968).
12. I. Marshall and A. Todd, *Trans. Faraday Soc.*, **49**, 67 (1953).
13. H. A. Pohl, *J. Am. Chem. Soc.*, **73**, 5560 (1951).
14. H. A. McMahon, *J. Chem. Engr. Data*, **4**, 57 (1959).
15. L. H. Buxbaum, *Angew. Chem., Int. Ed. Engl.*, **7**, (3), 182 (1968).
16. F. B. Marcotte, D. Campbell, J. A. Cleaveland, and D. T. Turner, *J. Poly. Sci. A*, **5**, 481 (1967).
17. J. G. Pacifici and J. M. Straley, *J. Poly. Sci., B*, **7**, 7 (1969).
18. J. H. Lemons, S. K. Kakar, and D. M. Cates, *Am. Dyst. Reptr.*, **55**, P76 (1966).
19. O. B. Edgar and R. Hill, *J. Poly. Sci.*, **8**, 1 (1952).
20. P. J. Flory, *J. Chem. Phys.* **17**, 223 (1949).
21. Saichi Morimoto, "Manmade Fibers," eds., H. F. Mark, S. M. Atlas, and E. Cernia, Wiley-Interscience, New York, 1968, Vol. 3.
22. U. S. Patent 2,895,986, DuPont.
23. B. C. M. Dorset, *Text. Mfgr.*, **92**, 241 (1966).
24. British Patent 734,416, British Celanese, Ltd.; U.S. Patent 3,065,207, DuPont; U.S. Patent 2,647,104, DuPont; Japan Patent 4595(66), Teijin, Ltd.

25. U.S. Patent 3,324,085, Union Carbide.
26. F. Guetbauer, E. Proksch, and H. Bildstein, *Oesterr. Chem. -Ztg.*, **67** (2), 35 (1966).
27. U.S. Patent 3,274,295, Dow Chemical Co.
28. British Patent 1,014,101.
29. Japan Patent 19,708(65), Toyo Rayon.
30. British Patent 1,040,730, I.C.I.
31. Japan. Patent 19,474(65), Kurashiki Rayon.
32. G. R. Ward, *Am. Dyest. Reptr.*, **44**, 220 (1955).
33. U.S. Patent 3,245,905, Eastman Kodak.
34. E. I. Valko and G. C. Tesoro, *Text. Res. J.*, **29**, 21 (1959).
35. U.S. Patent 3,242,117, Millmaster Onyx Corp.; British Patent 1,037,444, Hercules, Inc.
36. M. Seidel, *Tertilveredlung*, **2** (6), 356 (1967).
37. J. Kermit Campbell, *Am. Dyest. Reptr.*, **57**, 75 (1968).
38. U.S. Patent 3,379,564, DuPont.
39. Japan Patent 661(67), Toyo Rayon.
40. R. W. Singleton and P. A. C. Cook, *Text. Res. J.*, **39**, 43 (1969).
41. British Patent 981,539, Geigy.
42. U.S. Patent 3,308,095, DuPont.
43. British Patent 1,126,979, Ciba, Ltd.
44. British Patent 1,138,944, I.C.I.
45. Netherlands Application 6,606,892, DuPont.
46. British Patent 1,033,999, I.C.I.
47. O. Oldenroth, *Chemiefasern*, **17** (9), 726 (1967).
48. Ryuichiro Tsuzuki and Noboru Yabuuchi, *Am. Dyest. Reptr.*, **57**, 472 (1968).
49. U.S. Patent 3,420,697, Allied Chem.
50. French Patent 1,437,500, Minn. Mining and Mfg. Co.
51. U.S. Patent 3,412,175, DuPont.
52. U.S. Patent 3,377,249, Deering Milliken.
53. E. M. Perry, *Am. Dyest. Reptr.*, **57**, 405 (1968).
54. Wilson A. Reeves, John V. Beninate, Rita M. Perkins, and George L. Drake, Jr., *Am. Dyest. Reptr.*, **57**, 1053 (1968); *Chem. and Eng. News*, 40 (Aug. 12, 1968).
55. U.S. Patent 3,410,927, Monsanto.
56. C. E. Miles, H. L. Vandersall, and J. W. Lyons, paper presented Div. Org. Coatings and Plastics, A.C.S. San Francisco, Calif. 1968. See *Modern Plastics*, **45** (10), 87 (1968).
57. U.S. Patent 3,440,222, Stauffer Chemical.

Chapter 8 **POLYOLEFINS**

I. Introduction

The polyolefin fibers of importance today, polyethylene and poly-propylene, are linear polymers based on a paraffinic backbone. The interactions of their molecular weight, molecular weight distribution, degree of branching, and stereoregularity provide a classic example of the intricate relationships between molecular structure and fiber properties.

Of the two fibers, polypropylene has the more desirable properties and it is assuming a secure position in the synthetic fiber field. Poly-ethylene, having a lower melting point, is being used in cordage and netting where its resistance to microbial attack is advantageous.

K. Ziegler and G. Natta(1) discovered a catalyst system which polymerized α-olefins by an anionic mechanism. This catalyst system was unique in that it could control the stereoregularity of the mono-mer as it is added to the growing polymer chain. This control of regu-larity has produced polymer forms which Natta refers to as isotactic, syndiotactic, and atactic. The isotactic and syndiotactic forms are highly ordered and have the capacity to crystallize, while the atactic form is disordered and very little crystallinity can develop. Using poly-propylene as an example, isotacticity may be explained by describing the asymmetric carbon atoms in the hydrocarbon backbone as having the same steric configuration. The methyl groups attached to the asymmetric carbon atoms are described as being on the same side of the hydrocarbon chain.

$$\left(-CH_2-\underset{\underset{H}{|}}{\overset{\overset{CH_3}{|}}{C^*}}-\right)_n$$

For syndiotactic regularity, the methyl groups in polypropylene alternate in the following regular manner:

$$-CH_2-\underset{\underset{H}{|}}{\overset{\overset{CH_3}{|}}{C}}-CH_2-\underset{\underset{CH_3}{|}}{\overset{\overset{H}{|}}{C}}-CH_2-\underset{\underset{H_2}{|}}{\overset{\overset{CH_3}{|}}{C}}-$$

and in an atactic polymer the position of the methyl groups shows no regularity. It is the molecular symmetry associated with isotacticity in polypropylene which allows crystallinity to develop. In turn, crystallinity in an oriented fiber confers many useful physical properties. Stereoregularity is never perfect and the degree of isotacticity usually obtained will vary from 85–97%.

II. Polyolefin Fiber Properties

A. CHEMICAL

When polymerized, ethylene ($CH_2{=}CH_2$) and propylene ($CH_2{=}CHCH_3$) yield polymers of high molecular weights ranging from 100,000 to 2,000,000. These polyolefins are hydrocarbons and as such are very hydrophobic. Moisture absorption is virtually zero, the fibers are dimensionally stable with changing humidity and, therefore, their mechanical properties are unaffected by moisture. Because of their hydrophobicity, polyolefin fibers have excellent resistant to water-born stains and are easily cleaned. The water repellency of these fibers acts in a negative way by introducing problems in dyeing so that special treatments are needed in order to obtain colored fibers in an aqueous dyeing process. Although the fibers are generally resistant to organic solvents at room temperature, vegetable and mineral oils tend to be absorbed and swell the fiber at higher temperatures.

This is especially true of polyethylene. For example, polyethylene is soluble in some chlorinated hydrocarbons and aromatic solvents at 70–80°C. The polyolefins are resistant to a wide range of chemicals including acids and alkaline compounds but they are susceptible to oxidation unless adequately protected by stabilizers.

B. PHYSICAL

In the melt spinning of these fibers crystallization is so rapid that they cannot be quenched in an amorphous or very low crystalline state. However, as with all fibers, their physical properties will vary according to the size, type, and arrangement of the ordered polymer molecules in the fiber. The polyolefin fibers are lightweight with specific gravities ranging from 0.90–0.97 allowing for high bulk and coverage per unit weight of fiber. Linear polyethylene melts at 130–133°C and isotactic polypropylene melts at 160–175°C. Softening points of these fibers are approximately 10–15°C lower. The thermal properties limit considerably the areas where the fibers may be used (see Table I).

TABLE I

	Tenacity, g/den.	Elongation, %	Shrinkage 100°C, 20 min
Linear polyethylene	5–8	10–20	5–15%
Isotactic polypropylene	5–9	15–25	10–15

*a*Original length.

The quantitative relationships between the tensile strength and other mechanical properties of polypropylene fibers and molecular weight, molecular weight distribution, crystallinity, and orientation have been studied by Sheehan, Wellman, and Cole(2). The polypropylene samples studied had 94–98% isotacticity and weight-average molecular weights ranging from 245,000 to 2,400,000. Isotacticity was not a variable in the study. The equation which was developed shows that tensile strength is primarily a function of orientation and that, in comparison, the other three structural parameters make only a small contribution. The equation may be expressed in the following form:

$$\overline{TS} = 10^4[-0.16 + 0.76(1-\alpha)^{-1.5} - 0.18Q + 5.14 \times 10^{-6}Mw]$$

where $\overline{TS}$ is the tensile strength, α is the total molecular orientation, Q is the molecular weight distribution expression, and Mw means the weight-average molecular weight.

Polyethylene fibers tend to undergo creep over long periods of time that is somewhat excessive when compared with other manmade fibers. The extent of creep in polyethylene fibers increases with a decrease in chain branching, yet with less chain branching the tenacity increases. Thus, one physical property is improved at the expense of another. Polypropylene shows almost satisfactory creep characteristics, but it does not reach the standard set by polyamides and polyesters, particularly at elevated temperatures.

III. Radiation Reactions

A. POLYETHYLENE

Free radicals are produced in polyethylene when it is exposed to radiation from high-energy sources. These free radicals are able to react with a methylene group [—CH_2—] forming either inter- or intramolecular crosslinks. The extent and nature of the crosslinking are dependent on the orientation and crystallinity of the solid state. If a vinyl monomer is present, then reaction at the free-radical site would result in a graft, producing a polymeric side chain or branch from the vinyl moiety. The variety of high-energy sources which have been considered include x rays, γ rays, accelerated electrons and β particles from nuclear disintegrations, protons, deuterons, α particles, and neutrons. The utilization of these energies continues to grow as the economies in processing techniques become more practical. In polyethylene, radiation treatment induces crosslinking and prevents "collapse" at temperatures above the crystalline melting point. The temperature of the polymer during exposure to the energy source is an important parameter and controls the crosslinking efficiency as well as further crystallization and sperulitic growth. There are three reactions which occur when polyethylene is exposed to high-energy sources: (1) crosslinking, which has already been described; (2) chain scission; and (3) unsaturation(3). The presence of oxygen allows the formation of hydroperoxides as one of the intermediate products in each of these reactions. Further study has shown also that crosslinking can take place between layers of polyethylene single crystals if the lamellae are in close contact, and intra crosslinks occur in folded molecules in compacted lamellae(4).

B. POLYPROPYLENE

The structure of polypropylene $(-CH_2-C(H)(CH_3)-)$ with its vulnerable tertiary hydrogen introduces additional problems when exposed to high-energy sources. Temperature and environment (O_2, N_2, H_2O) influence the reaction paths. Radiation in air produces carbonyls and hydroxyls as well as hydroperoxides, whereas radiation in vacuum produces double bonds of the vinylidene type, and the product behaves as if it were crosslinked(5). When polypropylene was irradiated at 20°C, Milinchuk and coworkers(6) found more free radicals in the crystalline polymer than in the amorphous polymer. The location of the free radicals was reversed when irradiation was carried out at −195°C.

The physical properties of polypropylene are not particularly enhanced from exposure to high-energy sources since chain scission and crosslinking occur at about equal rates. Odian and Bernstein(7) and Sobue(8) both report a ratio of crosslinking to scission of 0.75–0.9:1, while Salovey and Dammont(9) found the reverse to be true. Thus, radiation of polypropylene induces chain scission reactions which are seldom desirable for improving many of the properties necessary for a textile fiber. The application of radiation will be discussed further when considering the chemistry of grafting on to polypropylene fibers.

IV. Degradation and Stabilization

Polyolefins may also be degraded by sources of energy other than radiation, inducing reactions best described in the following manner:

(1) *Thermal*—decomposition of compounds containing weak bonds.
(2) *Photochemical*—reactions in which the absorption of visible or uv light gives the molecule sufficient energy to break a chemical bond. The wavelength of light must correspond to an absorption band of the molecule.
(3) *Oxidation–reduction*—reactions that involve the gain or loss of an electron.

Oxidation reactions have been studied extensively with particular emphasis on polypropylene. The formation of hydroperoxides and free radicals are the initial steps in degradation sequences. Some of these have been summarized by Reich and Stivala(10).

$$RH + O_2 \rightarrow R\cdot + HO_2$$
$$R\cdot + O_2 \rightarrow RO_2\cdot$$
$$RO_2\cdot + RH \rightarrow RO_2R + H\cdot \text{ (one R may be H)}$$
$$RO_2\cdot + RH = \text{inactive products}$$
$$RO_2R \rightarrow 2RO\cdot$$
$$RO_2R \rightarrow \text{inactive nonvolatile products}$$
$$RO\cdot \rightarrow R'\cdot + \text{volatile products (such as } CO, CO_2, H_2O)$$
$$R'\cdot + RH \rightarrow R\cdot + R'H$$
$$R'\cdot + O_2 \rightarrow \text{less reactive products}$$

The absorption of oxygen, whether atomic or molecular, is dependent upon the temperature(11). Temperature is therefore an important factor in controlling the rate of degradation, and probably the mechanisms as well. Another influential factor is the morphology of the polymer which controls accessibility. The oxygen uptake is proportional to the amorphous region and with oxidation there is scission of chains in the disordered region. Scission of the constrained chains allows greater chain mobility resulting in further crystallization(12). In studying the relationships between crystallinity and oxidation, it was found that branching can nullify the effects of chain scission. Even if the amorphous portion does not vary appreciably, it has been shown that as the degree of branching increases so does the oxidation rate (13). In polyolefin oxidation this effect of branching can be attributed to steric factors which tend to control and weaken intra- and intermolecular forces and cause shielding. Also, in polyethylene, branching produces tertiary carbons whose singular hydrogen atom is more readily removed by radical reactions than are those on secondary carbon atoms. The rate of many oxidation reactions are dependent on the diffusion of oxygen into the sample. The diffusion, and consequently the rate of oxidation, are temperature depentent(14). However, there are experimental limits within which the diffusion rate is not affected appreciably by temperature.

Although, as catalyst contaminants, metals possess the ability to accelerate oxidative degradation, they also may intensify the effect of an antioxidant. The work by Rysharez and Balaban(15), Meltzer and coworkers(16), and particularly Hansen and coworkers(17) should be referred to for degradation mechanisms of polypropylene and the roles of metal contaminants. One proposed reaction scheme for oxidative degradation is as follows:

$$ROOH + Co^{2+} \longrightarrow RO\cdot + HO^- + Co^{3+}$$
$$ROOH + Co^{3+} \longrightarrow RO_2\cdot + H^+ + Co^{2+}$$

Thus, the overall equation represents a bimolecular decomposition of a hydroperoxide.

$$2ROOH \longrightarrow RO\cdot + RO_2\cdot + H_2O$$

The degradation products resulting from the oxidation of poly-olefins are mentioned throughout the literature. Products which have been observed include water, formaldehyde, acetaldehyde, acetone, methanol, hydrogen, hydrogen peroxide, carbon monoxide, and carbon dioxide. The mechanisms by which these products are produced are complex. Chien, Vandenberg, and Jabloner have studied the structure of polyprolylene hydroperoxides and their thermal decomposition (*18*). These workers found considerable intramolecular hydrogen bonding of the hydroperoxide groups by infrared spectroscopy. The chemical modification of such a polymer lead them to suggest that all the polypropylene hydroperoxide groups are present in sequences of lengths of two and greater. Thermal decomposition of the hydro-peroxide is believed to consist of two consecutive reactions, of which the initial, faster reaction is 60 times that of the slower process(*19*). This fast reaction was suppressed by 2,6-di-tert-butyl-p-cresol which reacted with two oxygenated radicals produced in the beginning of the initial reaction. The role of the scavenger begins as follows:

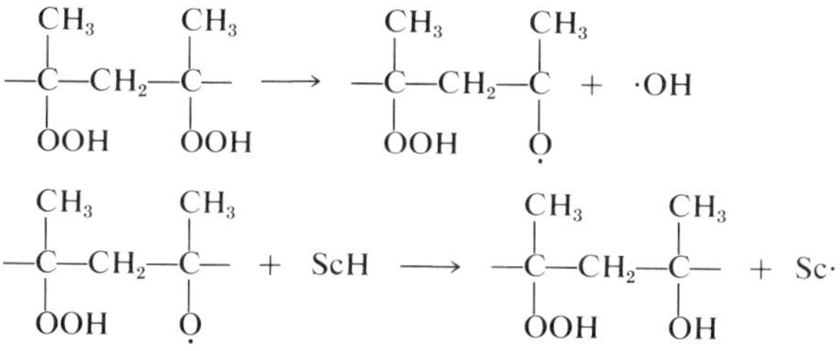

It is in this way that stabilization is achieved and the stabilizer is con-sumed. Chien and Jabloner also include a detailed description of their reaction sequence for the decomposition of polypropylene hydro-peroxide via a variety of species which are postulated to exist. These include both the neighboring and isolated hydroperoxides.

The literature, as one would suspect, overlaps in the areas of oxidative, thermal, and uv stabilization. Oxidation, as has been dis-cussed, can be due to oxygen and heat, oxygen and uv light, traces of metal catalyst with oxygen, etc. Light stability is obtained with aromatic ketones, such as o-hydroxybenzophenones, hydroxyphenyl-benzotriazoles, aromatic thiophosphorus compounds, salicylates, benzoquinones, benzonilides, organotin, and nickel compounds, and other metal salts which frequently act as an aid to dyeability. Antioxidants are usually phenolic derivatives, and many of these are capable of complexing with catalyst residues as are tartaric acid, amides, and malonamides. Thermal stability is also improved by phenolic derivatives, phosphites, and phosphates. Certain thio compounds, when used in conjunction with phenolic stabilizers, show a synergistic effect and this combination has become an accepted stabilization system.

It is believed that sulfur compounds stabilize polyolefins because of their ability to decompose hydroperoxides into nonradical products. One class of antioxidants that has attained commercial importance are the thiobisphenols which can function both as radical terminators and as peroxide decomposers. In general, thioethers are less effective than disulfides in peroxide decomposition while the thiobisphenols are probably most effective as radical terminators. In polypropylene, however, the thioether, dilauryl thiodipropionate with a phenolic, is a very effective synergistic stabilizer. Despite this it does appear that the organosulfur compounds must first undergo some oxidation to an active product such as a sulfonic or sulfinic acid which is the active entity. The activity of the compound is not evident until substantial oxidation has occured. Hawkins and Saulter(20) suggest that the sulfur componds both decompose peroxides and regenerate chain terminators in antioxidant combinations, thus accounting for the high level of synergism observed.

Stabilizers are generally applied to the polypropylene pellets by a "dusting" technique or by a solvent technique before extrusion into fibers. These stabilizers also may be added during polymerization to produce a "master batch" of pellets which can then be mixed in the desired proportion with unstabilized pellets. The quantities required can vary from 0.001% to as much as 5%, based on the weight of the polypropylene.

Table II gives a few of the recent patents in the area of stabilization from several countries.

V.   Fiber Modifications

A. CROSSLINKING AND GRAFTING

The chemical modification of olefin fibers is economically prohibitive at present, but much research has been conducted in this area which will probably be utilized eventually. A number of problems have

TABLE II
PATENTS RELATING TO STABILIZATION

U.S. Patents	
Compound	Number
Aromatic ketones	3,098,842, DuPont
Nickel salts of aromatic hydroxycarbonyl compd.	3,098,863, DuPont
	3,244,688, Ethyl Corp.

$R^1 = C_{1-12}$ alkyl $R^3 = C_{1-2}$ alkyl
$R^2 = C_{3-12}$ $R^4 = C_{3-12}$
   α-branched alkyl α-branched alkyl
$Q = -CHO$ or NO_2 $R = H, C_{1-12}$ alkyl,
                              C_{7-12} aralkyl
                            $Z = H, C_{1-12}$ alkyl,
                                C_{6-12} aryl,
                                C_{7-12} aralkyl

Compound	Number
Zn dialkyldithiocarbamate and either (a) an alkylated phenol, (b) an alkylidenebisphenol, or (c) a trihydroxyphenylketone	3,249,583, Eastman
Complex organocitrate and neutral sulfur cpd., i.e., dilauryl thiodipropionate	3,251,792, Carlisle Chemical Works
Mono or polyhydric phenol + organic phosphite + thiopropionate	3,255,136, Argus Chem. Corp.
Organic thiophosphite, i.e., trilauryl thiophosphite	3,256,237, Sun Oil
SiO_2 or TiO_2 bonded with phenolic groups	3,259,603, Bell Telephone
$2'$-MeC$_6$H$_4$COC$_6$H$_3$(OH)$_2$-2,4	3,260,771, General Aniline & Film Corp.
Org. thiophosphate ester and γ radiation of the blend	3,261,804, W. R. Grace & Co.

TABLE II (Contd.)

U.S. Patents	
Compound	Number
Thiophosphorus Compound	3,264,257, Stauffer Chem. Co.

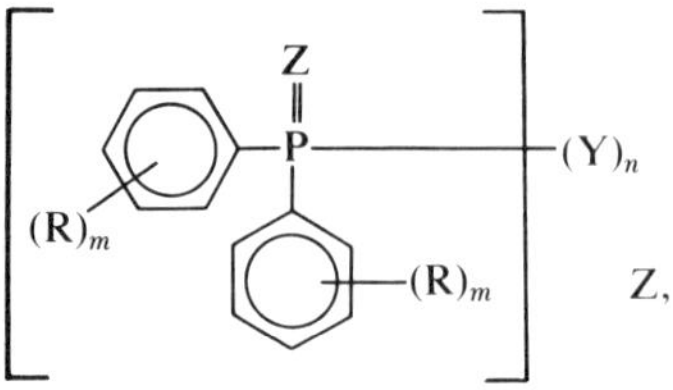

R = H, Cl, OH, or lower alkyl or lower alkoxy

Z, Y = S,O

Compound	Number
Complex chromium salt and a phenolic	3,268,476, Hercules Powder
2,6-Di-tert-butyl-4-methylphenol and dilauryl thiodipropionate	3,271,185, Phillips Petroleum
Chlorinated biphenyl or triphenyl to a stabilized poly-α-olefin	3,277,046, Johnson and Johnson
Di-β-naphthyl-p-phenylenediamine	3,282,889, FMC Corporation
$[RR'P(S)S]_2M$ R,R' = C_{2-18} alkyl, cycloalkyl, M = Zn, Ni, Pb, Cd, Ca, Cu, Sn, Mn	3,293,208, American cyanamid
Nickel organophosphites	3,395,112, Argus Chem.
9,10-Boroxaphenanthrene compounds	3,404,123, Mobil Oil Corp.
Diester of thiodipropionic acid and 3-hydroxy-2,2,4-trimethylpentyl isobutyrate	3,408,386, Eastman

British Patents	
Nickel additives	858,890; 914,336, Ferro Corp. 948,501–4, Sun Oil & Am. Visc. 945,050; 956,102; 958,830; 965,054, Hercules Powder 965,199, American Cyanamid

TABLE II (Contd.)

British Patents	
Compound	Number
Range of sulfonyl and thio compounds including phenylaryl sulfonates, sulfonamido -benzophenones, -diphenylsulfones, saccharins, dithio-oxamides, and thiophenols	910,766; 910,880; 911,065; 916,184, Montecatini
Hindered phenols, tertiary amines, sulfur compounds	949,394; 953,447–8, Farbwerke Hoechst 957,986, Geigy
β-cyanoethyl compounds	885,113, Montecatini
Compounds containing O-hydroxybenzophenone with alkoxy group C_{8-23}	918,464, DuPont
O-hydroxydiazobenzene	924,763, Farbwerke Hoechst
Nickel acetylacetonate	929,259, Montecatini
Nickel salts of aromatic o-hydroxycarbonyl compounds	943,937, Shell International
Zinc stearate (basic) and 2,2′-dihydroxy-4-octyloxybenzophenone	954,387, I.C.I.
	958,167, Geigy

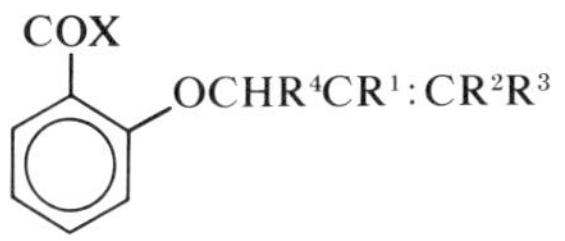

COX

OCHR⁴CR¹:CR²R³

2-Allyloxy-4-butoxybenzophenone

Aromatic sulfoxides or sulfones	958,830, Hercules Powder
Aromatic ketones	959,742, I.C.I.
Boron Compounds of the formula $(RR^1BH)_2$	1,023,931, Vereinigte Glanzstoff-Fabriken
Nitriles, i.e., PhCN, CH_2:CHCN, etc.	1,027,953, Vereinigte Glanzstoff-Fabriken
Substituted triphenodioxazines and 1,2,4-trichloro-7-nitro-3H-isophenorazin-3-one	1,034,181, Eastman

TABLE II (Contd.)

British Patents	
Compound	Number
Substituted 1,4-benzoquinones	1,034,182, Eastman
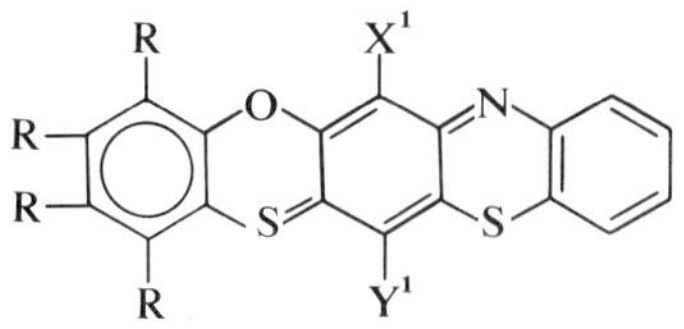	1,034,183, Eastman

Organic pigment

X^1, Y^1 = H, halogen
R = H, NH_2, NO_2, halogen, alkylamino, alkoxy, alkythio, phenyl

Episulfide polymer and phenolic antioxidant	1,039,120, Hercules Powder
Bis-methylene malonic acid compounds	1,115,596, Geigy

Japanese Patents	
Aromatic imides with other known antioxidants	13,056 (65), Toyo Koatsu Ind.
(1) light stabilizer, a salicylate,	13,057 (65),
(2) antioxidant, 4,4-thiobis	Mitsubishi Rayon
(2-methyl-6-tert-butylphenol)	
(3) 2nd antioxidant, dilauryl thiodipropionate	
Triorganotin compounds stabilize against light	13,259 (65),
such as tris(tributyltin) phosphate	Nitto Chem.
Triaryltin compounds such as	13,260 (65),
tris(diphenyl propyltin) phosphate	Nitto Chem.
Triorganotin compounds such as	13,261 (65),
tris(triphenyl propyltin) phosphate	Nitto Chem.
Hydroxybenzanilides such as	13,262 (65),
4-(lauroyloxy)-4-hydroxybenzanilide	Toyo Spinning
3,4-βz(OH)$C_6H_3O(CH_2)_m X(CH_2)_n$Me	17,135 (65),
$m = 2$–20	Mitsubishi Rayon
$n = 0$–20	
X = CO, CO_2, COS, S, SO, SO_2	
Disubstituted tartaric acid amides or	5389 (66),
malonamides	Sumitomo Chem.
Aryl distannates such as	6755 (66),
1,1,2,2-tetrabutyl-1,2-diphenyldistannane	Nitto Chem.

TABLE II (Contd.)

Japanese Patents	
Compound	Number
Substituted hydroxy naphthoic acid and esters	6756 (66), Shibata Rubber Ind.
Known stabilizers with malic acid	17,229 (66), Mitsubishi Petrochem.

Netherlands Applications	
Sodium stearate and 1,3,5-trimethyl-2,4,6-tris(3,5-di-tert-butyl)-4-(hydroxybenzyl) benzene	6,412,637, Shell International
Trisphenol such as 2,6-bis(2-hydroxy-3-tert-butyl-5-methylbenzyl)-4-methylphenol	6,504,241, Sun Oil
RN-(R″OH)[[R′N(R″OH)]$_n$R]	6,506,063, Universal Oil Products
(R$_2$N)$_2$P-(:O)SR′	6,506,303, Stauffer Chem.
Mixed ester of thiodipropionic acid	6,510,518, American Cyanamid
2,4-Dinonylphenol and 2,2′-methylene bis(4,6-dinonylphenol) with conventional stabilizer system	6,510,638, American Cyanamid
Nickel phenolate of 2,2′-sulfonylbis (p-tert-butylphenol), C$_{18}$H$_{37}$N MeP(O) (NH)$_2$ and 2,2′-isopropylidenebis (p-nonylphenol)	6,512,333, Farbwerke Hoechst
2-(2′-Hydroxy 5′-tert-butylphenol) chlorobenzotriazole	6,512,334, Farbwerke Hoechst
Phosphoric acid derivative.	6,602,632, I.C.I.

$C_{12}H_{25}OP(OR)O(CH_2)_nO-$

6,605,945, I.C.I.

appeared which are germane to the development of chemically modified polyolefins. Considerable degradation occurs in crosslinking polyolefins with high-energy radiation techniques, although this modification process is possible under vacuum conditions with uv light. Under
uv light, reaction is accelerated by the presence of inorganic substances, principally TiO_2 and Al_2O_3, but, unfortunately, it is still slow,
requiring 10–60 h(*21*). In general, the use of radiation is not a satisfactory method for modifying polyolefin fiber properties through the
introduction of crosslinks. A second method for property modification
is graft polymerization. As intimated in discussing degradation, there
are several ways to produce reactive sites in polyolefins, particularly
in polypropylene. One way is with a high-energy source such as γ rays
from a cobalt source, or even uv light can be used to produce free radicals. Chemically, an oxidizer such as tertiary benzoyl peroxide or an
O_2/O_3 gaseous mixture is adequate. The monomers which have been
used in grafting reactions include:

    (1) Acrylonitrile
    (2) Styrene
    (3) Divinylbenzene
    (4) Vinyl chloride
    (5) Vinyl acetate
    (6) Vinylidene chloride
    (7) Vinyl sulfonic acids and salts
    (8) Acrylic acid
    (9) Acrylates
  (10) Methacrylates
  (11) Vinyl pyridines
  (12) Vinyl pyrrolidinones

Some of these grafts improve physical properties such as strength,
elongation, and heat resistance; others improve adhesive properties
and wettability and the nitrogen-containing monomers and sulfonic
acids improve dyeability with acid and basic dyes respectively. Table
III lists some of the recent work using the monomers listed above.
An extensive survey of grafting has been prepared by Battaerd and
Tregear (*22*).

    The examples listed in Table III also represent the methods avail-
able for the synthesis of graft copolymers with polypropylene as the
backbone. Radiation techniques are used in several ways: (1) to form
an active site on a preformed macromolecule in the presence of a poly-
merizable monomer; (2) to form a peroxidized polymer in the presence

TABLE III
GRAFTING ON POLYPROPYLENE

Monomer	Initiator	Reference
Acrylonitrile	Radiation	Japan 18,329 (67)
	Radiation, Co source	S. L. Dobretsov, A. I. Kurilenko, V. A. Temnikovskii, and V. L. Karpov, "Probl. Fiz.-Khim. Mekh. Voloknistykh Poristykh Dispersnykh Strukt. Mater., Mater. Konf.," Riga, 1965, p. 551
	Free radical	Neth. (65) 16,717, Atlass S. M.
	uv light	K. Yahikozawa and T. Suda, *J. Fac. Text. Sci. Technol. Shinshu Univ. Scr. B.* (42), (1965); E. P. Danilov and A. I. Kurilenko, *Vysokomol. Soedin*, **9**, 2679 (1967)
Styrene	γ radiation	A. I. Kurilenko, E. P. Danilov, and V. A. Temnikovskii, *Vysokomol. Soedin*, **8**, 2024 (1966)
	Radiation (Paper II)	V. A. Temnikovskii, A. I. Kurilenko, and S. L. Dobretsov, *Mekh. Polim.*, **1967** (1), 167
	Radiation (Paper III)	*ibid.*, **1967** (6), 963
		A. I. Kurilenko, L. B. Aleksandrova, and L. B. Smetanina, "Fiz-Khim. Mekh. Orientirovannyk Stekloplast Sb. Rab. Konf.," Moscow, 1965, p. 90
Divinylbenzene	Bz_2O_2 or O_2O_3	J. Barton, *Khim. Volokna*, **1967** (3), 33
Vinyl chloride	Preradiation	K. Hayakawa and K. Kawase,
Vinylidene chloride	Co source in air	*J. Poly. Sci. PtA-1*, **5**, 439 (1967)
Vinyl acetate		A. I. Kurilenko and V. I. Glukhov, "Probl. Fiz-Khim. Mekh. Voloknistykl Poristykh Dispersynykh Strukt. Mater. Knof.," Riga, 1965, p. 605
Vinyl sulfonic acids or salts (Na, K, NH_4) + styrene or acrylic	Peroxidize PP in air and heat (120°C)	Japan 793 (67)
Vinyl	Form peroxide on olefin	U.S. 3,322,661, Chisso Corp.
Vinyl	O_2 under pressure elevated temperature	U.S. 3,288,739, Cook Paint and Varnish
Vinyl chloride	Radiation	A. V. Vlasov, Yu. P. Kudryavtsev, L. I. Malokhova, A. M. Sladkov, B. L. Tsetlin, and M. V. Shablygin, *Vysokomol Soedin*, *Ser. B*, **10** (2), 97 (1968).

TABLE III (Contd.)

Monomer	Initiator	Reference
Vinyl or vinylidene cpd Acrylic acid	Hydroperoxidize olefin Formation of hydroperoxide	Brit. 1,038,726, Chisso Corp. J. J. Wu, Z. A. Rogovin, and A. A. Konkin, *Khim. Volokna*, (5), 18 (1961)
Acrylic acid	γ radiation	W. Isuji, T. Ikeda, Y. Kurokawa, and N. Nakatani, *Sen-i Gakkaishi*, **23** (7), 327 (1967)
Acrylic acid	Co source, preradiated	W. Isuji, T. Ikeda, and Y. Kurokawa, *Bull. Inst. Chem. Res., Kyoto Univ.*, **45** (1), 87 (1967)
Acrylates Methyl methacrylate	Tesla coil H_2 atm peroxide formed on contact with air	T. Kakurai, K. Kaeriyama, and T. Noguchi, *Chem. High Polys* (*Tokyo*), **23** (254), 426 (1966) Brit. 1,032,505, Dow Chemical

$$CH_2{=}C-\overset{\overset{\displaystyle O}{\|}}{C}{-}(OCH_2CH_2)X$$

$$|$$
$$R$$

R=H, CH₃
X = halogen, OH, OR, etc.

Monomer	Initiator	Reference
Methyl methacrylate	O_2/O_3 mix, N_2H_2 sealed ampules	J. Vsiansky, and J. Jurak, *Przegl. Wlok.*, **21** (2), 74 (1967)
Alkyl methacrylate	γ radiation	K. Kawase, and K. Hayakawa, *Radiat. Res.*, **30** (1), 116 (1967)
Methyl methacrylate	Air oxidized at 60–70°C for 10 h	K. S. Minsker, and I. Z. Shapiro, *Trudy po Khim. i Khim. Tekhnol.*, **3**, 609 (1960)
Vinyl azo cpds.	Homo and graft produced these are polymerizable dyes – cellulose and PP	K. Uno, Y. Iwakura, M. Makita, and T. Ninomiya, *J. Poly. Sci.*, *PtA-1*, **5**, 23/1 (1967)
1-Phthalimido-1, 3 butadiene	Thermal mastication-roll mill, was melt spun at 275°C accepts acid dyes	A. Terada, *J. Appl. Poly. Sci.*, **12**, 35 (1968)
N-vinylpyrroli-dinone (N-vinyl lactam)	Radiation, uv may also be used with O_2/O_3 at 100°C for 2 min	U.S. 3,049,507, Dow
2-Vinyl pyridine or dimethylaminoethyl acrylate	As above	U.S. 3,049,508, Dow
2-Vinyl pyridine	Peroxidized in air tert. Bu_2O_2 vapor	E. Beati, S. Toffano, and S. Severini, *Chim. Ind.* (*Milan*), **45** (6), 690 (1963)
2-Vinyl pyridine or vinyl acetate, acrylic and methacrylic acid, etc.	Radiation	A. Robalewski and W. Zielinski *Polimery*, **9** (7–8), 310 (1964)
Acrylonitrile, vinyl chloride or 4-vinyl pyridine	5–15% C_4–C_8 boron alkyl in presence of 1–30/mole boron	U.S. 3,141,862, Esso
2-Methyl-5-vinyl pyridine	Oxygenated PP + $FeSO_4$/ MeOH. This was blended with "undenatured" PP for spinning	Japan 8927 (67) Mitsubishi Rayon

TABLE III (Contd.)

Monomer	Initiator	Reference
2-Methyl-5-vinyl pyridine + other vinyls	Irradiation	Japan 9836 (62)
Acrylonitrile, acrylic acid or lower esters	NO_2, N_2O_4 to produce reactive site. Treated polymer is then reacted with monomer at between 50°C and the boiling point of the monomer	U.S. 2,987,501 Dow Chemical

of oxygen which can then be used for a subsequent reaction for initiating the polymerization of a monomer; (3) preirradiation in vacuum to form trapped free radicals whose lifetime is prolonged at lower temperatures; and (4) radiation of an intimate mixture of two polymers producing two polymeric free radicals leading to a graft or interlinking of two different polymers. In grafting reactions, however, there is a possibility of chain-transfer reactions occurring which results in the production of some homopolymer. These reactions can be characterized by a chain-transfer constant. This constant represents the ratio of the velocity constant for transfer of chains to their growth constant. Other examples of grafting included in Table III are diazotization and degradation which produce free radicals by thermal and mechanical techniques.

The influence of structure is important in grafting. The initial rate is affected by the amorphous areas, their size and orientation, as well as segment mobility of the chains at various temperatures(*23*). Diffusion rates of monomers do not affect the chain growth of the graft but do affect its termination. Measurements of the activation energy of diffusion in various media sometimes suggest that the energy required for diffusion is used primarily in overcoming polymer–polymer interactions. In the case of polypropylene it has been shown that the structure of the diffusing molecule may have a profound effect on the activation energy(*24*). However, there are no simple correlations, for the diffusion of a molecule follows a tortuous path through amorphous areas whose structure can be very influential in the control of the graft molecular weight and dispersity.

Chain-transfer reactions are another very common method for synthesizing graft copolymers. By its very nature, the graft formed is in admixture with some homopolymer. The difference between a chain-transfer reaction, a polymeric radical transfer reaction, and a polymeric radical initiated graft has not always been made clear in the literature. This has been treated in detail by Ceresa(25) and by Burlant and Hoffman(26). Today the distinction between chain transfer and polymeric radical is usually made on the basis of kinetics or initiator choice. Hydroperoxide formation is a third method of producing grafted polyolefins. These may be formed at temperatures of 70–80°C with air, oxygen, or peroxy chemicals(27).

B. COSPINNING

Cospinning of polymers is of growing importance in the fiber industry if the patent literature is any indicator. The properties of polypropylene fibers can be modified in this manner to give crimp, dyeability, and oxidative and thermal stability. On the other hand, the cospinning approach to modification frequently leads to difficulties, since the second polymeric component may act simply as a diluent and a plasticizer.

In cospinning polypropylene with polyethylene, the polyethylene seems to act as a plasticizer and the best compatibility appears to be 75% polyethylene–25% polypropylene(28). Other reports of this polymeric blend, while not agreeing on the most desirable polymer contents, show that the fibers contain continuous polypropylene and polyethylene fibrils(29). The resulting fibers usually show abnormal behavior in their mechanical properties. Crimped polypropylene fibers can be produced by cospinning two different types of molten polypropylene. The polypropylenes should vary in molecular weight distribution but they also should have a similar melt viscosity. Fibers containing such polymers can be produced from spinnerets as a side-by-side composite filament or as a sheath core composite (30).

Many other polymers are cospun with polypropylene for the principal objective of improving and modifying dyeability. The polymers vary from such common ones as polystyrene and polyacrylates to particular dye acceptor types such as polyamides, polyimines, polytriazoles, and polypyrrolidones. Some of the more recent patents in this field are listed in Tables IV and V. The cospinning approach to polypropylene fiber modification is at present most useful for producing crimped or crimpable fibers and for improved dyeability. Frequently these properties are achieved at the expense of a decrease in another desirable physical property(31). Regardless, cospinning is an area where future progress may be anticipated for many melt-spun fibers.

TABLE IV
MODIFICATION OF POLYPROPYLENE BY COSPINNING

Description	Reference
1–20% wt. uncured epoxy resin made from	U.S. 3,013,998,
(a) epihalohydrin and a subst. bisphenol, hydroquinone, Na sulfide and aniline;	Montecatini
(b) aliphatic acid esters of condensates of epihalohydrin and bisphenol; and	
(c) styrenated fatty-acid esters of condensates of epihalohydrin with bisphenol	
5–50% plasticized vinyl chloride resin	U.S. 3,046,237, Dublon, Inc.
PVC 95%, 5% PE, 5% chlorinated PE	U.S. 3,085,082, Monsanto Chem.
Polystyrene, poly-3-methyl butene,polyhexene-1, polyisobutylene-1, polymethylmethacrylate	U.S. 3,121,070, Eastman
1–~ 25% wt. copolymer of ethylene and ethylenically unsaturated ester of a fatty acid	U.S. 3,163,492, Hercules Powder
Polyaminotriazoles, polytriazoles, polyethyleneimines, polyamides, epoxy resin, and dyeing in dye bath containing a naphthoic acid	U.S. 3,184,281, Asahi Kasei Kogyo Kabushiki Kaisha
Modified polyester (glycol 2–12 carbons, aliphatic dicarboxylic acid, 16–32 carbons, heterocyclic glycols, 1% chain terminator)	U.S. 3,223,752, Monsanto
1–30% copolymer ethylene and propylene and vinyl alcohol; an acetalization product between copolymer and aldehyde: and a urethanization product between said copolymer and urea.	U.S. 3,226,455, Kurashiki Rayon
$R + R'(-CH_2)_m-$ phenylene napthehalene, cyclohexylene, pyridylene and a group connecting $2(CH_2)_m$ groups through a sulfonic acid	U.S. 3,236,918, Asahi Kasei Kogyo Kabushiki Kaisha

1–20%

polytriazole

X, X' = H, CH_3, C_2H_5, phenyl, etc. = benzylamino, amino, pyrrolyl etc.

Polyhydroxyether	U.S. 3,297,784, Union Carbide
0.1–20% alkylene glycols and dicarboxylic acids	U.S. 3,312,755, Montecatini
2–20% N-vinyl polymers i.e. 80 parts N-vinyl pyrrolidone and 20 parts dimethylaminoethyl methacrylate	U.S. 3,316,328, J. J. Press
Unsaturated polyesters	U.S. 3,317,633, J. P. Stevens and Co.

TABLE IV (Contd.)

Description	Reference
2–20% N-vinyl polymers – described as hydrophilic, nonsoluble, fusible, and partially substituted polymeric composition such as N-vinyl methyl oxazolidinone, etc.	U.S. 3,337,651, J. J. Press
Linear polyamide and linear sulfonated polyamide	U.S. 3,328,484, Societe Rhodiaceta
1–25% polyamide	U.S. 3,331,888, Montecatini Edison

$$R-\overset{O}{\overset{\|}{O}}C-(CH_2)_n-N\overset{\overset{H_2\ H_2}{C-C}}{\underset{\underset{H_2\ H_2}{C-C}}{\quad\quad}}N-(CH_2)_n\ \ \overset{O}{C}-OR + \text{diamine aliphatic, aromatic and}$$

heterocyclic

Description	Reference
2–20% of (50–95% hydrophilic, nonsoluble in polyolefins fusible copolymer, N-vinyl pyrrolidone/vinyl acetate and 5–50% lower M.W. polar organic, i.e., alkyl esters of polyhydroxy compounds, fatty amides, etc.)	U.S. 3,337,652, J. J. Press
Silica gel 0.1–5%	U.S. 3,355,416, Mobile Oil
PP with nitrogen containing polymer dyed with Lewis acids in bath	U.S. 3,361,843, Uniroyal
Polyphenylene oxide + 1–10% polyolefin	U.S. 3,361,851, General Electric
1–25% basic nitrogen polycondensation product from dihalo alkane or dihalo alkanal with a bis-secondary diamine	U.S. 3,363,030, Montecatini
Hydrophobic nonsoluble in olefin – polyethylene glycol resin, CMC, sulfonated vinyl hydrocarbon polymers etc. and lower M.W., etc. – alkyl esters of polyhydroxy compounds, ethers, amines, phosphoric acids, etc.	U.S. 3,375,213, J. J. Press
PP and copolymer of ethylene and acrylyl compounds of the formula	U.S. 3,388,190, Union Carbide

$$CH_2\!=\!\underset{}{\overset{R}{C}}-\overset{O}{C}-R'\quad R\!=\!H,CH_3\quad R'\!=\!N\overset{R''}{\underset{R''}{\diagup}}$$

$R'' = H$, alkyl

Description	Reference
Poly-vinyl pyridine, -amine, -amide	Belg. 669.085, U.S. Rubber
An aftertreatment with an acid reagent	Neth. 6,507,660, U.S. Rubber
Polyvinylpyridine; aftertreatment with Lewis acid	Belg. 627.798, U.S. Rubber; Brit. 1.033.088, U.S. Rubber

TABLE IV (Contd.)

Description	Reference
Poly(2-vinylpyridine); aftertreatment with ethylene oxide, dried, then H_2SO_4 treated	Belg. 611,524, Montecatini
Polyvinylpyridines or polyacrylates poly(2-vinylpyridine)	Brit. 879,198, Montecatini
Condensation product with amino groups	Japan 3,292 (65), Montecatini
Spin copolymer of N-vinylpyrrolidone and methylacrylate	Japan 27,636 (64), Nitto Spinning
Copolymer of diallylamine and a vinyl pyridine	Japan 17,139 (65) Nitto Spinning
Copolymer of a vinyl pyridine–styrene with a poly(oxy-ethylene or oxypropylene) added	Japan 9,742 (65), Mitsubishi Rayon
Similar	Japan 934 (65), Mitsubishi Rayon
Polyamide resin and ethylene vinyl acetate copolymer	Brit. 1,111,210, Chemcell
Crystalline vinylpyridine polymers	Brit. 992,012, Montecatini
Poly(2-vinylpyridine) + antioxidant and uv light absorber	Brit. 923,407, Montecatini
Stereoregular polyvinylpyridine	Japan 15,824 (65),
Treating polyolefin/polymeric basic nitrogen mixture with tris or bis-(1(2-methyl) aziridinyl) phosphoxide	Brit. 1,009,907, Montecatini
Treating fibers of polyolefin/polyvinylpyridine mixture with a halogen in liquid or gaseous state	Ger. 1,216,481, Montecatini
Polyethylenimine followed by halogenation and cyanamidation	Japan 05,555 (68), Asahi Chem.
Polyester and Zn stearate	Japan 14,578 (66), Toyo Rayon
Blend polymer or copolymer of 4-vinylpyridine and styrene	Japan 06,533 (68); 06,534 (68), Mitsubishi Rayon
Add copolymer of N-methylmethacrylamide, styrene, N-isopropylacrylamide vinyl chloride, etc.	German 1,254,287, Eastman
Chlorinated or chlorosulfonated product of a polyolefin, uncured epoxy resin, a condensate or a basic compound and formaldehyde containing one unit of	Japan 265 (68); 266 (68), Nihon Orimono Koko K.K.

and a compound containing an amino group with at least one $=N-H$

TABLE V
ADDITIVES FOR DYEABILITY

Description	Reference
Addn. of melamines	Japan 17.140 (65), Nitto Boseki
Aftertreat fibers with cyanuric chloride or subsequently with NH_3 or amines	Japan 6235 (66), Nisshin Spinning
Add polyesters, polyurethanes, polyureas, polycarbonates, poly(oxyalkylenes)	Japan 14,578 (66), Toyo Rayon
Polyvinyl acetate added	Brit. 975,918, Toyo Rayon
Chlorinated polypropylene	Japan 11.171 (65), Toyo Rayon
Acrylic or methacrylic esters	U.S. 3,156,743, Eastman
Simple amines such as octadecylamine, hexadecylamine	Brit. 985,937, Nippon Rayon
Addition of copolymer of ethylene and N-vinylsuccinimide. N-vinylpyrrolidone, or acrylamide	Brit. 995,802, Union Carbide
Nitrogenous polymer made with piperazine and a mono or dichlorinated aliphatic or cycloaliphatic hydrocarbon and then epichlorohydrin	Brit. 1,006,581, Montecatini
Terpolymer of styrene, acrylonitrile, and a polar monomer	Japan 17.136 (65), Nitto Spin.
Nitrogenous condensate of epichlorohydrin and piperazine spun in with aftertreatment of polyvinyl alcohol and a lower aldehyde	Brit. 1,009,616, Montecatini
Brit. 1,009,616 but final treatment with piperazine or derivative	Brit. 1,009,661. Montecatini
Blend polyamides or N-alkylated polyamides	Brit. 1,005,725, BASF
Blend copolymers containing S-triazine rings	Brit. 1,003,005, I.C.I.
Blend polyesters	Can. 717,587, Eastman
Modified polyesters	Can. 717,586, Eastman
Sulfonate-modified polyesters	Brit. 998,688, Union Carbide
Sulfonate-modified polyamides	Brit. 1,014,913. Soc. Rhodiaceta
Addition of diphenylether sulfonates	U.S. 3,207,725, Dow Chemical
Add polymers having p-C_6H_4O and —CONH linkages	Japan 17,764 (65), Nitto Boseki
Polycondensate containing basic N used in aftertreatment	U.S. 3.245.751. Montecatini
Add copolymer of C_2H_4 and acrylic ester, $CH:CRCO R$ which has been treated with an amine	Neth. Appl. 6,510,533, Sumitomo Chem. and Toyobo
Blend polyester of ethylene terephthalate and adipate	Japan 7.449 (66), Toyo Rayon

C. Metal Complexes

The addition of a second polymer to polypropylene has been discussed and this technique can improve and control the dyeability of the fiber. Likewise, the addition of metal salts or complexes can produce similar modification and improved dyeability. The metals most frequently referred to are aluminum, zinc, nickel, cobalt, cadmium, chromium, calcium, barium, and magnesium. One of the basic considerations in this work is to incorporate into the fiber, by prior blending or coating of the pellets, the metal in an organo form. This is because compatibility or solubility in the melt is required in order to have the dye-attracting moiety either uniformly dispersed throughout the fiber or else uniformly dispersed along the fiber's sheath.

The function of the metal in the fiber is to fix the dye by chelation. The bond formed through chelation is usually very strong and can produce good dye fastness. Problems of leveling during dyeing may be experienced because of the rapid exhaustion of the dye bath. Fortunately, chelation does not usually change the effect of a metal-containing uv stabilizer.

Table VI lists some of the recent patent literature on metal salts for dyeability. Fordemwalt *et al.*(32) have published a general review of the subject which gives an excellent insight into the problems of dyeability. Although extensive research has been devoted to this subject, nickel compounds appear to be the most satisfactory. In addition, color may be given to polyolefin fibers by the blending of pigments with the resin before spinning. This method of coloring fibers has been evaluated in most man-made fibers and has found commercial usage in several.

TABLE VI

DYEABILITY OF POLYPROPYLENE BY ADDITION OF METAL COMPOUNDS

Description	Reference
Treat article with organoaluminum, i.e., triethyl aluminum and hydrolyze	U.S. 3,371,982, Hercules
Werner chrome complexes of amino acids	Ger. 1,255,306, Gagliardi
Zn, Cd, Al, Cr, Co, Ni salt of $(HOCH_2CH_2)_2NCS_2H$ or morpholino-dithiocarbamic acid	Japan 9267 (67), Yoshimura Oil Chem.

TABLE VI (Contd.)

Description	Reference
Metal xanthates	Japan 11,485 (67), Toyo Spinning
Metal salts of fatty acids	Japan 788 (67), Mitsubishi Rayon
Metal salt of a carboxylic acid	Japan 20,106 (66), Mitsubishi Rayon
Ex.: $C_{18}H_{37}NHCOCH_2CH_2CO_2H + (iso\text{-}PrO)_3Al$	Japan 3107 (67), Mitsubishi Rayon
Cu, Zn, Cd, Co, Ni salts of C_{4-18} fatty acid	Belg. 691,918, Hercules
Esters of silicic acid, i.e., tetrahexyl orthosilicate	Brit. 1,101,609, Dynamit Nobel
Zn salts of carboxylic acids and salts of Schiff base	Japan 03,969 (68), Ube Nitto Chemical Industry
Al compounds coordinated with carboxylic acids	Japan 09,398 (68), Daiichi Lace Mfg.
1,2,4-Triazole complexes of Zn, Ni, Cr, Al	Fr. Addn. 90,064, Asahi Chem. Ind. Japan 19,813 (66), Mitsubishi Rayon

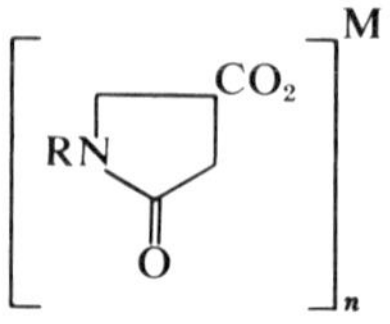

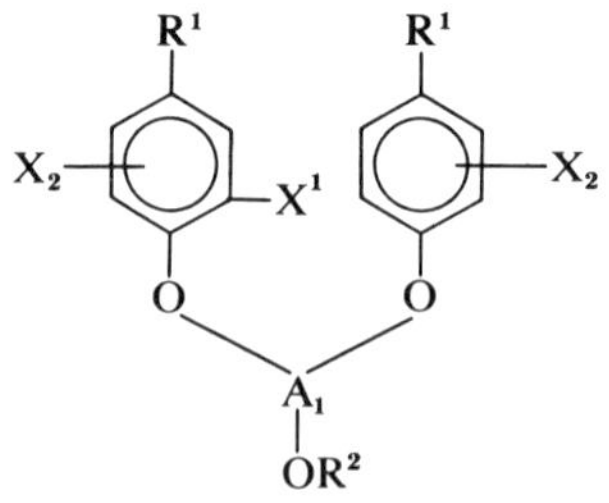

M = Ba, Ca, Zn, Al, Cr, Co, Ni
R = 1–30C atoms alkyl
    6–30C atoms aralkyl

Japan 19,819 (66), Mitsubishi Rayon

| Similar to above structure | Japan 19,820 (66), Mitsubishi Rayon |
| Metal complex of $RCO_2M(OR')_2$ | Japan 19,985 (66), Mitsubishi Rayon |

R = alkyl of < 12 C atoms
R' = alkyl
M = Mg, Zn, Cd, Al

TABLE VI (Contd.)

Description	Reference
Treated with Ni derivatives of an α amino acid of the formula RCH(NHR')COOH R = 1–8 C atoms R' = H or phenyl	U.S. 3,299,030, Esso R. and E.
Blending Ni derivative of a diamine R'R²C(NHR³)CH₂NH₂ R'R² and R³ = HC₁₋₁₈ alkyl or aryl groups	U.S. 3,284,428, Esso R. and E.
Metallic salt and a dibasic acid or its anhydride	Japan 14,661 (66), Toyo Rayon
Zn, Cd, Hg complexes of β-diketones or hydroxy ketones	Japan 14,750 (66), Toyo Rayon
Mg, Zn, Ni, Al complex compounds of	Japan 19,814 (66), Mitsubishi Rayon

Treated with Ni derivatives of an
  α amino acid of the formula
$RCH(NHR')COOH$ $R = 1-8$ C atoms
                $R' = H$ or phenyl
Blending Ni derivative of a diamine
  $R'R^2C(NHR^3)CH_2NH_2$
  $R'R^2$ and $R^3 = HC_{1-18}$ alkyl or aryl groups
Metallic salt and a dibasic acid or its
  anhydride
Zn, Cd, Hg complexes of β-diketones or
  hydroxy ketones
Mg, Zn, Ni, Al complex compounds of

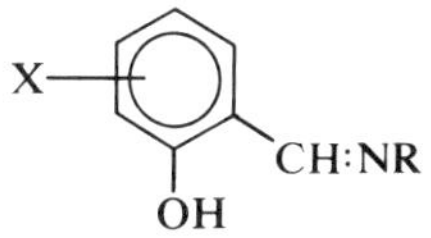

$R = $ alkyls of 5–30 C atoms
$X = $ H, alkyl, Cl, Br, I, NO₂, CN
Al compounds of

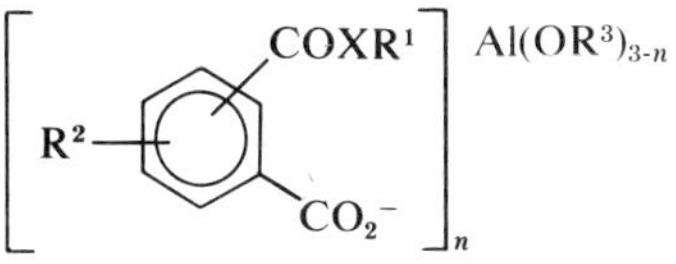

$R^1 = $ alkyl 4–30 C atoms
$R^2 = $ H or alkyl 1–30 C atoms
$X = $ NH or O
$R^3 = $ alkyl 1–20 C atoms
$n\ \ = 1, 2$ or 3

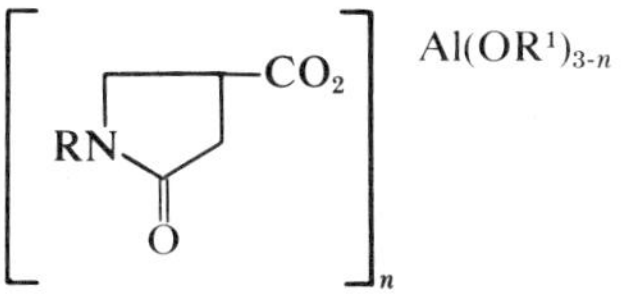

$R\ \ = $ alkyl 1–30 C atoms
$R^1 = $ alkyl 1–20 C atoms
$n\ \ = 1, 2,$ or 3

Japan 19,815 (66),
  Mitsubishi Rayon

Japan 19,816 (66),
  Mitsubishi Rayon

TABLE VI (Contd.)

Description	Reference

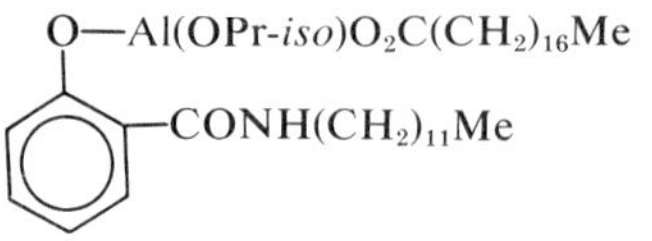

O—Al(OPr-*iso*)O$_2$C(CH$_2$)$_{16}$Me

⟨benzene ring⟩—CONH(CH$_2$)$_{11}$Me

Japan 22,588 (67),
Mitsubishi Rayon

with or without Zn, Ni salts of org. acids

Metal complexes of polyphenols (Al, Ni, Zn)

Japan 06,527 (68),
Yoshimura Oil Chem.

[R^{1}R^{2}R^{3}N(OH)CH$_2$CO$_2$]$_k$ M$_m$ or
[X[N(OH)R^{1}R^{2}CH$_2$CO$_2$]$_2$]$_p$M$_q$
  X = alkylene or phenylene
  M = Cu, Mg, Ca, Sr, Ba, Zn, Cd, Al, Sn, Pb, Ti,
    Cr, Mn, Fe, Co, Ni

Japan 7451 (66),
Yoshimura Oil Chem.

Add chelates of metals of Group IB or VIII and
compounds of MeCOCH$_2$COR', Bz$_2$, CH$_2$, etc.

Japan 6349 (66),
Toyo Rayon

Add Cu, Zn, Cd, Al, Cr, Co, Ni, salts of
RXC(O) CH$_2$SCH$_2$CO$_2$H

Japan 19,984 (66),
Yoshimura Fat Chem.

Metal salts, metal chelates such as Zn, Cd

French 1,418,743,
Montecatini

Metal-containing polyurethane, polyamide or
copolyurethane, copolyamide

Brit. 1,013,042,
I.C.I.

Metal salt of amino-triazole

French 1,393,883,
F. By.

Bis-(p-alkylphenol)sulfoxide or sulfone
all or part of phenolic H replaced by Cr

Brit. 996,109,
Hercules Powder

Salt or chelate of Ni, Co, Cr, and also
poly(vinylbutyral) or poly(methylmethacrylate)

Brit. 1,007,337,
Eastman

Complex acid such as phosphotungstic + amine
or metal salt and resulting precipitate blended

Japan 12,876 (65),
Mitsubishi Rayon

Metal complexes such as Cr and Al stearate or
acetylacetonate

Brit. 1,016,953,
Farbwerke Hoechst

Alkylsulfonic Al alkoxides

French 1,416,413,
Mitsubishi

Alkyl ester or alkali salt of alkylenediamine-N,N'-
tetraacetic acids aftertreatment with Ni(SCN)$_2$

Japan 6347 (66),
Nitto Boseki

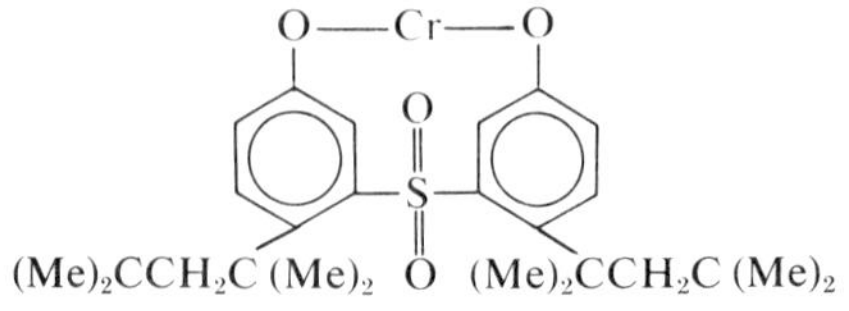

O——Cr——O

⟨structure with two benzene rings joined by sulfonyl group⟩

(Me)$_2$CCH$_2$C (Me)$_2$ O (Me)$_2$CCH$_2$C (Me)$_2$

Me = methyl

Brit. 996,109,
Hercules Powder

TABLE VI (Contd.)

Description	Reference
Sulfonates or low molecular weight polyester classes	U.S. 3,256,363, Union Carbide

MO_3S—⟨OC$_n$H$_{2n}$O⟩—⟨CO$_2$R, CO$_2$R⟩   X—⟨SO$_3$M⟩ RO$_2$CH$_{2n}$C$_n$ C$_n$H$_{2n}$CO$_2$R

M = metal

REFERENCES

1. G. Natta, *Mod. Plast.*, **34** (4), 169 (1956).
2. W. C. Sheehan, R. E. Wellman, and T. B. Cole, *Text. Res. J.*, **35**, 626 (1965).
3. R. M. Black and A. Charlesby, *International J. App. Radiations and Isotopes*, **7**, 126, (1959).
4. R. Salovey and A. Keller, *Bell System Tech. J.*, **40**, 1409 (1961); ibid., 1497 (1961); R. Salovey, *J. Poly. Sci.*, **61**, 463 (1962); R. Salovey and D. C. Bassett, *J. Appl. Phys.*, **35**, 3216 (1964).
5. H. Sobue, Y. Tajima, and Y. Tabata, *Kogyo Kagaku Zasshi*, **62**, 1777 (1959).
6. V. K. Milinchuk, S. Ya. Pshezhetsku, A. G. Kotov, V. I. Tupikov, and V. I. Tsivenko, *Vysokomol. Soedin.*, **5**, 71 (1963).
7. G. G. Odian and B. S. Bernstein, *Amer. Chem. Soc., Div. Polym. Chem., Preprints*, **4**, 393 (1963).
8. H. Sobue, Y. Tajima, and Y. Tabata, *Kogyo Kagaku Zasshi*, **62**, 1774 (1959).
9. R. Salovey and F. R. Dammont, *J. Poly. Sci., Part A*, **1**, 2155 (1963).
10. Leo Reich and S. S. Stivala, *Rev. Macromol. Chem.*, **1**, 249 (1966).
11. R. H. Hansen, Proc. Symp. Polypropylene Fibers, Sou. Res. Inst., Birmingham, Ala., 1964, p. 137.
12. F. H. Winslow, C. J. Aloisio, W. L. Hawkins, W. Matreyek, and S. Matsuoka, *Chem. and Ind. (London)*, **1963**, 533; F. H. Winslow and W. Matreyek, paper presented at Chicago meeting, ACS Division of Polym. Chem. Sept. 1964, **5**, p. 552; J. P. Luongo, *J. Poly. Sci.*, **B1**, 141 (1963).
13. A. H. Willbourn, *J. Poly. Sci.*, **34**, 569 (1959).
14. J. E. Wilson, *Ind. Eng. Chem.*, **47**, 2201 (1955); G. W. Blum, J. R. Shelton, and H. Winn, *Ind. Eng. Chem.*, **43**, 464 (1951); C. R. Boss and J. C. W. Chien, *J. Poly. Sci.*, *A-1*, **4**, 1543 (1966).
15. D. Rysharez and L. Balaban, *SPE Trans.*, **2**, 5 (1962).
16. T. H. Meltzer, J. J. Kelly, and R. N. Goldez, *J. Appl. Poly. Sci.*, **3**, 84 (1960).
17. R. H. Hansen, C. A. Russell, T. DeBendictis, W. M. Martin, and J. V. Pascale, *J. Poly. Sci.*, **2**, 587 (1964).
18. J. C. W. Chien, E. J. Vandenberg, and H. Jabloner, *J. Polym. Sci., Part A*, **6**, 381 (1968).
19. C. E. Boozer, G. S. Hammond, C. E. Hamilton, and J. N. Sen, *J. Am. Chem. Soc.*, **77**, 3233 (1955).
20. W. L. Hawkins and H. Saulter, *J. Poly. Sci., Part A*, **1**, 3499 (1965).

21. J. R. Hatton, J. B. Jackson, and R. G. J. Miller, *Polymer*, **8**, 411 (1967); C. Kujirai, S. Hashiya, H. Furuno, and N. Terada, *J. Poly. Sci., Part A-1*, **6**, 589 (1968).

22. H. A. J. Battaerd and G. W. Tregear, "Graft Copolymers," Wiley-Interscience, New York 1967, p. 257.

23. Louise Oder, *J. Poly. Sci., Part C* (22), 477 (1968).

24. R. A. Jackson, S. R. D. Oldland, and J. Pajaczkowski, *J. Appl. Poly. Sci.*, **12**, 1297 (1968).

25. R. J. Ceresa, "Block and Graft Copolymers," Butterworths, London, 1962.

26. W. J. Burlant and A. S. Hoffman, "Block and Graft Polymers," Reinhold, New York, 1960.

27. G. Natta, E. Beati, and F. Severini, *J. Poly. Sci.*, **34**, 685 (1959).

28. N. V. Mikhailov, E. F. Fainberg, V. O. Gorbacheva, and Ching-Hai Ch'eng, *Vysokomol. Soedin.*, **4**, 237 (1962).

29. Akio Nojiri, Haruo Morimoto, and Asamu Ishizuka, Kobunshi Kagaku, **24**, 250 (1967); T. V. Druzhinina, Yu. D. Andrichenko, and A. A. Konkin, *Khim. Volokna*, **1967**, 12.

30. British Patent 1,123,952, Asahi Chemical Co.; British Patent 1,049,933, I.C.I., Ltd.; Netherlands Application 6,605,271, I.C.I., Ltd.

31. D. V. Filbert, V. P. Murav'eva, Yu, K. Vasli'ev, and A. B. Pakshver, *Khim. Volokna* (3), 18 (1968).

32. F. Fordemwalt *et al., Am. Dyest. Reptr.*, **54**, 107 (1965).

Chapter 9 **SPANDEX**

I. Introduction

The Textile Fiber Products Identification Act designates Spandex as the generic term applicable to elastomeric fibers in which the fiber-forming substance consists of at least 85% of a segmented polyurethane. These elastic fibers are characterized by their high extensibility and their subsequent rapid and nearly complete recovery(*1, 2*). The term spandex does not include those fibers and yarns whose elasticity has been developed by physical and chemical treatments of textile fibers inducing structural changes and geometric deformations, such as the heat setting of a false-twisted yarn. The term spandex also does not include rubber threads. Spandex, as described by Gregg(*3*), achieves its elasticity from a combination of isocyanate chemistry and fiber technology.

II. Structure

A. CHEMICAL

The chemical structure of spandex polymers is complex and only general formulas will be used to represent their structure. Spandex polymers are copolymers derived from low molecular weight aliphatic polyesters or polyethers(4) with terminal hydroxyl and/or carboxyl groups generally coupled by reaction with diisocyanates(5). The coupling reaction builds molecular weight to a level sufficient for fiber formation. The polyester or polyether units are described as the "soft" segments, and the diisocyanate components as the "hard" segments. The principal characteristics of the soft segments are their flexibility, low molecular weight (800–3,000), melting points less than 50°C, and the presence of reactive terminal groups. The molecular weight of spandex polymers may be increased by coupling reactions in a variety of ways.

$$HO-R-OH + OCNR'NCO \longrightarrow H\left[O-R-O\overset{O}{\overset{\|}{C}}NH-R'-NH\overset{O}{\overset{\|}{C}}\right]_n-OROH$$

Coupled via urethane linkage

R = soft segment from polyester or polyether
R' = "hard" segment from diisocyanate

If an excess of diisocyanate is used, the polymer chains will be terminated by isocyanate groups(6). These may be further reacted with a diamine or water to increase molecular weight and to vary the structure. Such reactions introduce urea linkages into the polymer backbone.

$$OCN-polymer-NCO + H_2N-\square-NH_2 \longrightarrow$$

$$H_2N\left[\square-NH-\overset{O}{\overset{\|}{C}}HN-polymer-NH\overset{O}{\overset{\|}{C}}-NH\right]_n-\square-NH_2$$

Urea linkage

$$OCN-polymer-NCO + H_2O \longrightarrow \left[-polymer-NH\overset{O}{\overset{\|}{C}}-NH-\right]_n + CO_2$$

Thus, urethane linkages and urea linkages may be incorporated in the polymer structure as described above. The low molecular weight polyesters usually contain aliphatic dicarboxylic acids, such as adipic and appropriate diol(s). A number of diols have been mentioned in the patent literature including ethylene glycol, 1,4-butanediol, 1,6-hexanediol, p-bis(hydroxyethoxy)benzene, 1,3-propanediol, 1,2-propanediol, 1,3-dihydroxy-2,2-dimethylpropane, and phenyl–diethanolamine. When a polyether is used for the soft segment its molecular weight may be no more than 1000, as in the case of poly-tetramethylene glycol. Other soft segments may be polycaprolactone, polythioether, polyacetal, polyacetate, polyesteramide, or poly-oxyalkylenediamine. Diisocyanates usually used to couple these segments are methylene bis(4-phenyl isocyanate) and 2,4-toluene diisocyanate. Others mentioned in the literature include p-xylene diisocyanate, hexane diisocyanate, and $\alpha,\alpha,\alpha',\alpha'$-tetramethyl-p-xylene diisocyanate. The diamines may be ethylene diamine, hexamethylene diamine, 1,3-diaminopropane, 1,3-diaminocyclohexane, hydrazine, 2,5-dimethyl piperazine, and 4,4'-methylene bis(cyclohexylamine). Aside from the use of these compounds for linear chain extensions the polymer can undergo another reaction which further modifies its ultimate fiber properties. The urethane and urea groups in the polymer chains can be crosslinked with a diisocyanate. An allophanate-type linkage results from the crosslinking of two urethane groups whereas a buiret-type linkage results from crosslinking the urea groups. If both urea and urethane groups occur in the linear polymer, cross-linking principally involves the urea groups at temperatures around 90°C, and the urethane groups when crosslinking is conducted at about 140°C. This crosslinking or branching reaction increases molec-ular weight as well as the viscosity of the polymer solution and may lead to gel formation. If crosslinking and branching are desired, the reaction may be carried out after fiber formation either in the coagula-tion bath or during the application of surface finish. All of these reactions are graphically described in Fig. 1, and some of the patent literature is listed in Table I.

Other elastomer polymers which have been reported include a polycarbodiimide based on poly(tetramethylene)glycol(8), and a poly-

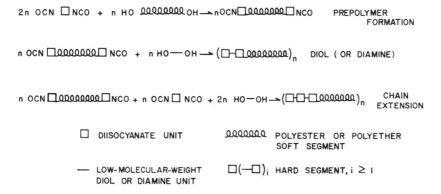

FIG. 1. Schematic equation for polyurethane formation, modified form of Gregg's descriptive equations(7).

TABLE I
RECENT PATENT LITERATURE ON FORMULATIONS FOR SPANDEX
FIBERS

Netherlands Application

291,290	Polythane Corp.
291,291	Polythane Corp.
6,501,927	I.C.I. Fibers, Ltd.
6,508,295	Eastman Kodak Co.
6,508,297	Eastman Kodak Co.
6,508,384	E.I. duPont de Nemours and Co.
6,510,297	Farbenfabriken Bayer A.G.
6,510,841	Globe Manufacturing Co.
6,513,521	Farbenfabriken Bayer A.G.
6,515,899	Farbenfabriken Bayer A.G.
6,516,154	Hans Peters
6,516,282	Monsanto Co.
6,600,989	United Elastic Co.
6,605,294	Farbenfabriken Bayer A.G.
6,608,798	E.I. duPont de Nemours and Co.
6,609,902	E.I. duPont de Nemours and Co.
6,612,988	Farbenfabriken Bayer A.G.

British

1,009,388	Mobay Chemical Co.
1,038,167	John V. Worthington
1,038,191	John V. Worthington
1,038,355	Thiokol Chemical Corp.
1,044,716	Upjohn Co.
1,057,505	K. H. Schmidt, E. Roethe

French

1,408,314	Societe' des Usines Chimiques Rhone-Poulenc
1,423,945	Societe' Rhodiaceta
1,433,347	Hokusin Chem. Industries, Co., Ltd.
1,441,388	Globe Manufacturing Co.
1,450,606	Societe' des Usines Chimiques Rhone-Poulenc
1,474,248	Fillatice S.P.A.
1,475,471	Upjohn Co.
1,522,611	Kurashiki Rayon Co., Ltd.
1,522,757	Monsanto Co.
1,534,257	Frederico Urgesi
1,534,536	I.C.I., Ltd.

United States

3,198,863	Thiokol Chemical Corp.
3,214,411	Mobay Chemical Co.
3,267,192	Polythane Corp.
3,271,346	Asahi Chem. Industry Co., Ltd.
3,294,752	E. I. duPont de Nemours and Co.
3,373,139	E. I. duPont de Nemours and Co.

carbonate derived from poly(tetramethylene)glycol, 4,4'-(2-norbornyl-dene) bis(2,6-dichlorophenol)(9). These polymers have good elasto-meric properties although there are no hydrogen bonds which are generally considered to be the "tie points" in a urethane–urea block polymer. In the polycarbodiimide, it has been suggested that there are transient crosslinks of the type, —N═C—N— serving as "tie points,"

$$\begin{array}{c} | \quad | \\ -N-C\!\equiv\!N- \end{array}$$

In the polycarbonate system, it may be that chain slippage is prevented by the very bulky hard segment. A similar explanation would account for the elasticity in a polyester that is a block copolymer prepared from dimethyl terephthalate and butanediol and polytetramethylene glycol. The block polybutylene terephthalate forms a high-melting hard segment which prevents undue chain slippage when the fibers are stretched. The high elongation is provided by the soft segment consisting of the condensation product from dimethyl terephthalate and the polytetramethylene glycol(10).

B. Fiber Formation

Fiber formation may be accomplished either by wet spinning, dry spinning, or melt spinning, or by variations of any of these established methods. Wet spinning is the extrusion of a solution of the polymer through orifices in the spinneret into a coagulation bath. In the wet spinning of spandex-type polymers, solvents such as N,N-dimethyl-formamide and N,N-dimethylacetamide are suitable. Also in the cate-gory of wet spinning there is the technique of reaction spinning in which the viscous isocyanate-terminated noncrosslinked prepolymer is extruded into a bath which contains "hardening chemicals." These chemicals include water, glycols, alcohols, and amines. Such fibers, when first coagulated, are weak, and exposure to the bath is for a short time to solidify the fiber surface only. Final hardening or curing is usually accomplished in a water wash bath during which the molecular weight is increased by reaction of the hardening chemicals with the isocyanate-terminated prepolymer, as shown in Fig. 1.

Dry-spinning methods also require that the polymer be dissolved in a solvent, but after extrusion through a spinneret, the newly formed filament is passed into a column or spinning tube which contains hot gas to drive off the solvent. Low-boiling solvents are preferred and dimethylformamide is most often used in dry spinning. Trichloroethane–formic acid and tetramethylene sulfone(11) can be used as solvents with certain polyurethanes.

Spandex fibers may be produced by a melt-spinning method provided a linear, thermoplastic polyurethane is used. Even so, there are problems associated with this method due to the reversibility of isocyanate reactions which will occur if the polymer is subjected to shearing conditions and prolonged exposures to high temperatures. This reversibility can lead to excessive crosslinks and gel formation. Because of such instability, melt spinning is the least desirable of the spinning methods and is probably not used commercially at present.

C. PHYSICAL

In discussing the chemical structure of spandex fibers, it was emphasized that the polymer consists of blocks of hard segments which are capable of strong interactions, leading possibly to crystallization. These hard segments alternate with the soft segments which provide the elastic and more flexible portions of the molecule. Since orientation in a spandex fiber would produce low elongations, increased stiffness and increased heat shrinkage there is no drawing operation after spinning, although some orientation naturally occurs in spinning. From measurements of mechanical properties and on the basis of structural studies theories have been proposed to explain the elastic behavior in terms of solid state structure. In the unstretched fiber, the molecules are pictured as being essentially in a state of random disorder with only slight orientation and no crystallinity. When the fiber is stretched, the soft or flexible segments of the molecule (which may be considered as randomly coiled or folded) are extended. The "hard" urea or urethane segments, which are stiff and subject to strong hydrogen-bonding interactions, are brought close together, attracting each other and acting as a crystal lattice, thus inhibiting further stretching. Removal of the stretching force allows local Brownian motion; this aids the flexible segments to resume their coiled configuration, and results in the length recovery of the fiber. Cooper and Tobolsky(12) studied the viscoelastic behavior of linear-segmented elastomers through the use of modulus–temperature experiments. The low-temperature properties of the block copolymer are governed by the nature of the flexible segments. These segments are the major portion of the polymer and therefore determine the major glass transition temperature (T_g). The hard segments in conjunction with these soft segments provide a broad temperature range of enhanced rubbery modulus.

The effect of cyclic stretching of the fiber permits a gradual interaction and buildup of more and more crystalline segments. These crystalline segments, being due to hydrogen bonding, increase the fibers resistance to stretch, or better, increase its tensile modulus. This change in modulus is affected also by such factors as time, temperature, degree of deformation and solvents. If the fiber contains chemical crosslinks, there is an additional mechanism that resists stretching and these crosslinks are not very sensitive to time and temperature.

III. Fiber Properties

A. CHEMICAL

The chemical structure of spandex fibers, while somewhat varied among manufacturers, does contain some common characteristics such as the urethane and urea groups derived from the diisocyanate. These groups are believed to be responsible for the discolorations, usually yellow, which are produced by hypochlorite and chlorite bleaches, nitrous oxide fumes, and uv light. However, in general, the fibers are more resistant to oxidation reactions than natural rubber. Peroxide and perborate appear to be safe bleaching agents for the spandex fibers. If the spandex fiber is based on a polyester, it is more subject to hydrolysis than one based on a polyether. Thus, resistance to dilute alkali and dilute or milk acids can be good but some of the fibers are degraded by strong alkalis and acids at elevated temperatures. These conditions of hydrolysis cause considerable loss in the tenacity and modulus of the fiber. The fibers are resistant to most solvents and oils, although glycol-based oils may cause some deterioration in certain fibers. Exposure to hot water in some instances reduces the strength of spandex fibers. Spandex dyes satisfactorily and shows a greater affinity for certain dyestuffs than nylon.

B. PHYSICAL

The general physical properties of spandex fibers are given in Table II. Some of the physical properties have quite broad ranges and do not fully describe the spandex fiber, particularly the property of elastic recovery. An elastic fiber is characterized by a high elongation and a low tensile modulus coupled with a high degree and rate of recovery from deformation. Recovery is usually studied by cyclic stressing of the yarn between 0% elongation and some upper elongation such as the breaking elongation less 100%. Usually upon continued cycling,

TABLE II

Property	Range
Tenacity, g/den.	0.55–0.9
Elongation, %	400–700
Elastic recovery, %	$\geqq$ 97 at 50–600%
Specific gravity	1.2–1.4
Water absorbency, %	0.3–1.3
Thermal properties	sticks at 350–400°F
	melts at 450–> 550°F

no significant change in properties occurs after five to eight cycles. After extension of the yarn, holding for 15 sec, and relaxing, a retraction curve is obtained which shows the restoring force or power of the yarn. The difference between the work required to extend the yarn and the work performed by the yarn in retracting is termed its hysteresis, and a low hysteresis is desired for elastomeric fibers. Decay of stress in the fiber at a fixed elongation is also an important property, particularly in the construction of fabrics and sizing of garments. Along with this, the property referred to as "set" should be carefully evaluated. This is the amount of permanent growth or creep that the elastic fiber undergoes after being elongated. A large amount of set is usually accompanied by a loss of modulus after long use or cyclic stressing. The set experienced in the commercial spandex fibers ranges between 10–25%, although experimental fibers have shown a range between 0–100%(13). The set can be reduced by increasing the isocyanate content during production of the polymer and by increasing the cure, or by stretching the fiber during a heat treatment. Wilson(14) has shown that thermal treatment has a pronounced effect on the stress–strain behavior of several commercial spandex fibers, relative to their set dimensions, and on the retractive force at a given extension. Their elastic recoveries were improved, their breaking extensions increased and their strengths decreased only slightly after thermal treatment.

IV.   Stabilizers and Finishes

A. Stabilization

As is evident from the chemistry of the spandex fibers, their properties are determined and controlled during the polymerization and subsequent spinning processes. With all the synthetic fibers, the producer is continually striving to improve fiber properties, particularly stability toward environmental conditions. Thus, the spandex fibers, which are most commonly used in blends with hard fibers such as cotton, rayon, polyester, and nylon in support clothing and sportswear, require stability toward uv light, perspiration, and combustion gases, as well as washability and resistance toward dry-cleaning solvents.

Considerable research has been devoted to the problem of discoloration and yellowing that results from uv light, chlorine and its compounds, and combustion gases. Improved stability toward uv light can be obtained by aftertreatment and/or curing with 2,4,6-trialkylphenol copolymer and an organic phosphite(*15*), formaldehyde(*16*), dialkylhydrazidophenols(*17*), diaryldiamines(*18*), special benzotriazoles(*19*), nickel complexes of substituted benzophenones (*20*), and carbodiimides(*21*). Yellowing due to gases requires somewhat different chemicals, whose function, as those for uv stabilization, may act as "blocking agents" at the remaining reactive or vulnerable sites in the polymer. Such compounds as phenylthioureas(*22*), acetic anhydride(*23*), hydroxytriamines(*24*), triphenylphosphite(*25*), and organic chlorides(*26*) are reported in recent patent literature to give protection.

The processing of spandex fibers has several problems associated with the formulation of the finish or lubricants which need to be applied(*27*). Frequently there is only limited compatibility with antistatic and emulsifying agents, and in view of the extreme porosity of some spandex fibers, there may be excessive finish pickup. Yet, lubricants are required immediately after spinning in order to control or prevent the fusion of the sticky, newly formed filaments. These surface finishes allow handling and decrease static electricity buildup in further processing. Cohen and Turer(*27*) evaluated lubricants by studying their effect on elongation in a simple laboratory test. Their results indicated that suitable finishes could contain white mineral oil, metal stearates, triglycerides, hydrocarbon waxes, silicone derivatives, and some poly(oxyethylene) derivatives. Compounds such as fatty acids, fatty esters, and low molecular weight glycols, and compounds with phenyl groups and certain nitrogen derivatives were unsuitable. These observations appear to be confirmed in the patent literature with examples of finish formulations including a mineral oil(*28*), a textile oil(*29*), magnesium stearate soaps, polydimethylsiloxane(*30*), liquid paraffin(*31*), and polyethyleneoxide esters of stearic acid.

REFERENCES

1. German Patent 826,641, Bayer Farbenfabriken.
2. U.S. Patent 2,650,212, Bayer Farbenfabriken.
3. Robert A. Gregg, "Kirk-Othmer Encyclopedia of Chemical Technology," Wiley-Interscience, New York, 1969, Vol. 18, p. 614.
4. U.S. Patent 2,929,803, du Pont.
5. U.S. Patent 2,625,535, Wingfoot Corp.
6. U.S. Patent 2,621,166, Farbenfabriken Bayer.
7. "Kirk-Othmer Encyclopedia of Chemical Technology," Wiley-Interscience, New York, 1969, Vol. 18, p. 625.
8. T. W. Campbell and K. C. Smeltz, *J. Org. Chem.*, **28**, 2069 (1963).
9. K. P. Perry, W. J. Jackson, Jr., and J. R. Caldwell, *J. Appl. Poly. Sci.*, **9**, 3451 (1965).
10. A. A. Nishimura and H. Komagata, *J. Macromol. Sci. (Chem.)*, **A1**, 617 (1967).
11. U.S. Patent 3,111,368, du Pont.
12. S. L. Cooper and A. V. Tobolsky, *Text. Res. J.*, **36**, 800 (1966).
13. R. A. Gregg, F. G. King, and M. W. Chappell, *Am. Dyes. Reptr.*, **53**, 334 (1964).
14. N. Wilson, *J. Text. Inst.*, **58**, 611 (1967).
15. Netherlands Application 6,508,298, Eastman Kodak Co.
16. British Patent 1,147,988, Courtaulds, Ltd.
17. French Patent 1,546,262, Farbenfabriken Bayer A.G.
18. Japanese Patent 11,680(67), Kurashiki Rayon Co., Ltd.
19. Netherlands Application 6,509,745, E. I. duPont de Nemours and Co.
20. French Patent 1,438,674, General Aniline and Film Corp.
21. German Patent 1,223,101, Farbenfabriken Bayer A.G.
22. Netherlands Application 6,502,388, Toyo Spinning Co., Ltd.
23. British Patent 1,095,922, Toyo Spinning Co., Ltd.
24. Belgian Patent 651,156, Badische Anilin- and Soda-Fabrik. A.G.
25. U.S. Patent 2,230,193, E. I. duPont de Nemours and Co.
26. U.S. Patent 3,462,297, Monsanto Co.
27. N. R. Cohen and Jack Turer, *Textile Bull.*, **92** (2), 36 (1966).
28. Netherlands Application 298,620, E. I. duPont de Nemours and Co.
29. U.S. Patent 3,277,000, E. I. duPont de Nemours and Co.
30. Netherlands Application 297,031, E. I. duPont de Nemours and Co.
31. French Patent 1,422,131, Kurashiki Rayon Co., Ltd.

Chapter 10 GLASS

I. Introduction

According to A.S.T.M. glass is an "inorganic product of fusion which is cooled to a rigid condition without crystallizing." When glass is extruded and produced in the form of a filament, it becomes a useful textile material capable of being used as a decorative, flexible fiber to fulfill many of todays consumer needs. At the same time, in view of its high strength and high modulus, glass fiber is used for industrial fabrics, and as a reinforcing material in composites. Production capacity of textile glass fiber is presently 588 million lb and it is predicted that a 34% increase in capacity will be achieved by late 1971 (*1*).

II. Structure

A. Chemical and Physical

The properties of different glasses depend on the dimensions and arrangement of ions in the silicate network. The major portion of the glasses consist of oxygen and the cations of silicon, boron, aluminum, calcium, magnesium, lead, barium, zinc, sodium, and potassium in varying combinations. Other cations such as lithium, phosphorus, berylium, zirconium, and titanium are used in special formulations.

The ionic radius of oxygen is 1.40 Å and that of silicon is 0.41 Å. Since the ionic radius of oxygen is the greater, the structure of glass could be considered to be oxygen ions grouped around cations. In the silicate lattice, silicon is surrounded by four oxygens which are arranged into a tetrahedron, with each of the four oxygens at a corner and identical with the corner of the adjoining tetrahedron, thus forming a continuous network. Yet a basic characteristic of glass structures is that this network in silicate glasses is irregular and aperiodic. Glasses are isotropic, and their structure cannot be expressed by a stoichiometric formula. Thus, polymeric groups of silicates contained in glass are not mutually equivalent. This aperiodic character of the structure of glass results in properties that are continuously changing with changing temperature and with no characteristic transitions. Glasses do not have a "melting point," and if cooled at a sufficiently fast rate they do not crystallize.

The structural elements in a crystal are arranged into a regular pattern but glass with its irregular network may be viewed as having holes or interstices in it. These holes are filled with cations which have relatively large ionic radii and small charges, and it is these cations which specifically modify the properties of the glass. The cations may be divided into two types: (1) those forming the network, such as silicon and boron, and (2) those found in the interstices where they modify the final properties, such as sodium, calcium, and others. Some of the elements used in glass are amphoteric and may enter the network or be found in the interstices. For example, one thinks of —Si—O—Si— linkages in glass, but in multicomponent compositions containing sodium, —Si—O—Na is formed. The presence of such monovalent cations in silica glass causes a breaking of the solid structure, increasing fusibility, decreasing viscosity, and reducing chemical and thermal stability.

In SiO_2, the silicon-to-oxygen ratio is $1:2$ or 0.5, and in Na_4SiO_4 it is $1:4$ or 0.25. Thus, as cations are put into silica glass, the ratio of silicon to oxygen decreases. "Soft glass" with greater quantities of nonbridging oxygens has a low Si/O ratio, and "hard glass" has a high ratio approaching 0.5. In a general manner, the Si/O ratio may be looked upon as a measure of polymerization of a glass. With a high ratio, there is a greater number of bridging oxygens and a high degree of polymerization, a high softening temperature and a low coefficient of thermal expansion.

Much of our present day understanding of the nature of glass is due to Zachariasen(2), who postulates several rules which govern the glass-forming ability of oxides:

(1) The oxygen atom is linked to not more than two metal cation atoms.
(2) The number of coordinating oxygen atoms surrounding the metal cation must be as small as possible, four or less.
(3) Polyhedra surrounding the cation share corners but not edges or faces with similar polyhedra.
(4) At least three corners of the polyhedra must be shared in order to form a three-dimensional network.

There are a variety of formulations for glasses which may be used in the production of glass fibers. Several general types whose compositions fall within general ranges are listed in Table I.

TABLE I
TYPES OF GLASS FOR FIBER PRODUCTION (3–5)

Type	General description	Possible composition range, %	
C	For use in filtrations	SiO_2	60–65
	of chemicals	Na_2O, K_2O	8–12
		MgO, CaO	15–20
		Al_2O_3	2–6
		B_2O_3	2–7
E	Alkali-free glass	SiO_2	52–56
	electrotechnical	B_2O_3	8–13
	equipment and fibers	Al_2O_3	12–16
	for reinforcement of	CaO	16–25
	plastics	MgO	0–6
		$Na_2O_1 K_2O$	0–1

B. FIBER FORMATION AND PROPERTIES

Glass is spun as continuous filament or as a staple fiber, and almost all continuous filament yarn is made from E glass and most staple fibers are prepared from a C-glass formulation. For both forms of the fiber, the manufacturing processes are identical in the making up of a batch of a given composition.

After a batch is melted to form the raw glass, the present-day process is usually to form the fibers directly from this melt. Originally the raw glass was formed into glass marbles which were subsequently fed into electric furnaces, remelted and formed into filaments. From the melt furnace the molten glass flows through orifices in the bottom of a platinum-alloy bushing and the continuous filaments are collected on high-speed winders. Winding speeds, orifice size, and glass viscosity are used to control fiber diameter. Staple fibers are spun as short filaments 8–15 in. long which are formed by jets of air that pull the fibers from the furnace and deposit them on a revolving drum.

Both continuous filament and staple fibers have a different nomenclature from other fibers (6). Fiber diameter and filaments per strand vary considerably and the industry offers a variety of diameters, filaments per strand, and strand counts. An E-glass fiber will have the following general properties:

Tenacity, g/den.
  dry   6.0–7.3
  wet   3.9–4.7
Elongation, %
  dry   3–4
  wet   2.5–3.5
Elastic recovery, %   100
Specific gravity   2.54
Moisture regain   none
Moisture absorbency up to 0.3% (surface)
Softening point, °F   1555

Glass fibers are resistant to most acids but are attacked by hydrofluoric, concentrated sulfuric, hydrochloric, and hot phosphoric acids. Hot solutions of weak bases and cold solutions of strong bases will attack glass causing deterioration of the fiber. Chemicals in general use do not attack glass and only the size on the surface of the fibers may be affected by organic solvents. Other glasses have been developed with even greater strength and a lower specific gravity, but these have not yet found their way into general commercial use.

III.   Surface Properties

A. PHYSICAL

An important feature of glass, which must be considered in conjunction with its chemistry, is the surface of the glass in fiber form. The fiber surface and the size which is applied can adsorb moisture even though glass fibers are not hydroscopic. The surface flaws affect the fiber's response to later chemical treatments as does the glass' chemical structure. Process control is more important probably in the spinning of glass fibers than in any other manmade fiber. For example, the viscosity of a glass while related to its composition also varies continuously with the temperature, and from controlling its flow through an orifice, through the drawing process and until it ceases to flow, the nature and uniformity of the surface is being determined(7).

In studying the failure behavior of aluminoboro silicate glass fibers, Kroenke(8) concluded that the structure is isotropic because the fibers follow the Weibull statistical theory of failure and the maximum principal stress theory of failure. However, the structure fixed in a glass fiber during its formation at high temperatures corresponds to a nonequilibrium state of the glass, and it would seem that the fiber would not be completely stabilized throughout its cross section(9). It has been found that there is a structure difference between the surface and the inner layers of the glass fiber. Etching techniques and uv spectra showed that the surface layer was more ordered than the structure of the inner fiber(10). In the subsequent applications of chemicals to glass-fiber surfaces, the order of the surface structure contributes to successful processing. It is certain that higher strength levels are obtained in fibers without defects, and low levels of strength are found in heat-treated glass fibers, probably because of the presence of surface microcracks produced by the further changes in its solid state(11). Thus, the method of forming the fiber influences its properties(12) and if surface flaws are prevented or minimized, strength remains high and fatigue failure is minimal(13). Hollinger and Plant(14) have subjected E-glass fibers to static fatigue tests in the high-stress region to clarify the role of atmospheric reactants in determining long-term strength. Analysis of their results led to the postulation of a distribution of flaws, at least some of which arise from chemical or physical segregation in the glass structure. They concluded that E-glass fibers could be improved by preventing the formation of flaws through better process control. This includes preventing airborne dust from hitting the hot fiber as well as control of phase separation within the glass structure.

B. CHEMICAL

Surface treatment of glass fibers usually vary with the ultimate performance requirements. In general, a "forming" size or binder is applied to the fiber immediately after it is formed, and usually contains a starch with a lubricant (usually cationic) or plasticizer in the formulation. A warp size is also frequently applied to the yarn in slashing to give added protection in weaving. These are normal warp sizes based on starch or polyvinyl alcohol. Such sizes must be capable of removal by pyrolysis during the weave-setting operation. Polyvinyl acetate is frequently used with a functional silane or methacrylate chrome complex. The size formulations can also include cellulose derivatives, gelatin, wetting agents, and emulsifiers in an aqueous solution. One such formulation includes starch, hydrogenated corn oil, polyoxyethylene sorbitan monooleate, tetramethylenepentamine distearamide, and polyethylene glycol (molecular weight 300)(*15*). A size of stearic acid in 1,1,1-trichloroethane may be used for protection from moisture and its degradation(*16*). Upon extrusion a size or binder such as an emulsion of polyvinyl acetate and $1,8\text{-}C_{10}H_6(CO_2Bu)_2$ has been claimed(*17*). Sizing can also be a two-step process involving a pad bath containing adipoyl chloride in water and a bath containing hexamethylenediamine in carbon tetrachloride(*18*).

With the size on the fibers, they can then be handled and woven. For further treatment of the fabric this size is removed before a finish is applied. Removal of the starch-based size is carried out by a heat process. It may be batch or continuous. The batch-heat process is a carefully controlled heat cycle reaching a temperature of about 600–700°C. This slowly burns off the organic material and produces a clean glass fabric with a slightly alkaline surface. It usually retains only about 40–60% of its original strength which is its major disadvantage. Otherwise, the heat process is simple, inexpensive, and uniformity can be maintained from batch to batch. A continuous desizing process involves passing fabric through a muffle furnace at about 1200°F with a treatment time ranging from 2–20 sec. Desizing by heat treatments also relieves internal stresses and induces crimp or weave set. Immediately upon cleaning, the fiber–fabric should always be given a finish to protect the surface(*19*).

Chemical cleaning to remove the size, although used very little in the industry, can involve a scour using detergents, enzymes, or sequestering agents. Depending on the nature of the cleaning bath, temperatures up to 100°C are employed. The chemically cleaned fabric usually retains as much as 0.10% organic material which is considered an acceptable level. This process is expensive and does not heat the fabric, but strength losses are much less(*20*). Combinations of these two cleaning processes may be found; for example, a trichloroethylene bath followed by a mild heat cleaning treatment, or a pretreatment with KNO_3, $RbNO_3$, or $CsNO_3$ followed by heating at 1000–1200°F for 1 h(*21*) appears very promising.

After cleaning the fabric a finish is applied and the type of finish will depend on the end use of the fabric. For decorative fabrics, the finish usually consists of acrylic copolymers, silane coupling agents, pigments, and "hand modifiers." This finish may be aftertreated with a stearate chromic chloride Werner complex for additional finish durability. It is important that the finish protect the fiber, give the required durability and have good pigment-holding properties.

A second type of binder is applied to yarns, roving, or fabric to be used for reinforcement. This binder must protect the fibers, be compatible with different types of resin, and give maximum strength. Glass fibers for composites require special finishes depending on their end use such as a silane coupling agent or chrome complex.

IV. Finishes

Some of the most important chemistry in finishes centers around the coupling agent which is the adhesive that forms the interface between the glass surface and the resin or polymer which will be applied at a later stage. It is this interface which determines the degree of bonding of the resin. Eakins(*22, 23*) has studied this bonding by examining the surface areas via electron photomicrographs, determining surface activity by titration, and measuring the physical properties of resin-bonded composites. Eakins as well as Tikhomirov(*24*) offer evidence for the importance of not just the polymer-fiber adhesion properties but of the importance of an oriented surface layer of the polymer which increases the mechanical strength.

Throughout all of these studies there is a concern about the porosity of the glass-fiber-reinforced laminate and the effect of water absorption on the glass–polymer bond. The porosity depends on the nature of the resin, type, surface pretreatment, and content of the glass, conditions of the lamination process, and the structure of the laminates (*25*). This debonding of the glass and resin may be described as a two-stage process. First, the resin swells by water absorption and this swelling eventually becomes large enough to produce a radical tensile stress at the fiber–resin interface. Secondly, irreversible hydrolysis of chemical bonds occurs, hastened by this stress, and cleavage occurs in the region that first becomes too weak to sustain the combined swelling and load stresses(*26*).

Lee(27) has developed an equation relating the glass temperature of a polymer to its wettability, and suggests that this parameter can be useful in predicting the adhesion of polymers to glass fibers at the interface. The importance of wettability is self-evident and would control the extent of the chemical reaction between the coupling agent and the polymer. In addition, the polarity and functionality of the polymer will add its influence to the system. The important role of wettability has been expressed in another manner by saying that the strength of the resin-to-glass bond may be limited by the chemical bond of coupling agent to glass rather than that of the coupling agent to the resin(28).

A. CHROMIUM COMPLEXES

Although the work of Englehardt and coworkers(29) has recently shown that the chromium complex, methacrylate chromic chloride

$$H_2C = C - CH_3$$

is only slightly superior as an adhesive, this coupling agent is designed to be compatible with the glass and organic polymers which are applied as a laminate or coating. Chromium complexes as coupling agents are used both on fibers or fabrics which are destined for resin impregnation and molding and on fabrics destined to be printed for use in a decorative manner. In this latter role the binder is often applied and the chrome complex, such as stearate chromichloride, is an aftertreatment(30). This complex was chosen on the theory that it could react with the Si—OH on the glass surface, and that the organic moiety would react or be compatible with resin. The chromium complexes are often called Werner complexes and as such are coordination compounds.

Theoretically these compounds contain a central atom or ion with a group of ions or molecules surrounding it. The charge on the coordinated species may be positive, negative, or zero, depending on the charge of the central atom and of the coordinated groups which are called ligands. The stability of these metal complexes, in general, increases if the central ion increases in charge, decreases in size, and increases in electron affinity. The platinum metals, because of their large polarizing ability, usually form the most stable metal complexes. The ligand's characteristics, which can affect dyeability, for example, are (1) basicity, (2) the number of metal chelate rings per ligand moiety, (3) the size of the chelate rings, (4) steric effects, (5) resonance effects, and (6) the ligand atom coordinating with the metal. As one would expect the most stable complex is also the least reactive or most inert. Since coordination compounds are formed as a result of acid–base reactions where the metal ion is the acid and the ligand is the base, the more basic the ligand the more stable, the more chelated the more stable, and the more double bonding in the chelate ring the more stable.

In the use of Werner chromium complexes as coupling agents for a glass resin bond, the following reaction may occur:

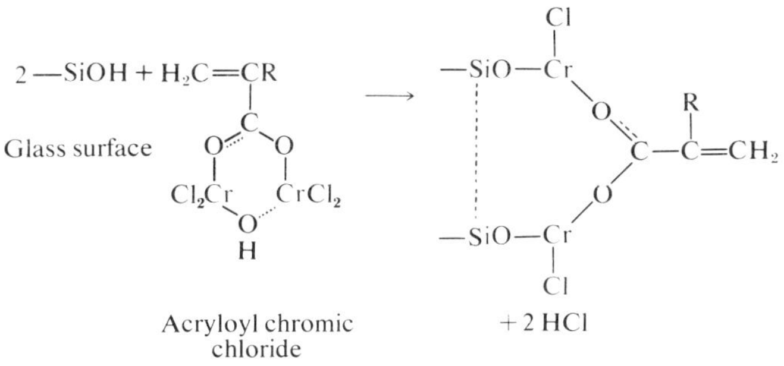

Acryloyl chromic chloride

$+ 2 HCl$

These acrylic-type chrome complexes have been used for a number of years(*31*) and have been very successful in the finishing of glass fabrics. Some of the more recent claims involving Werner chrome complexes are (1), a Cr (III) aminopropylate or glycine complex(*32*) for bonding glass fibers to elastomers such as neoprene and (2) as methacrylate chromic chloride used in conjunction with a telomeric acid as an anchoring agent. The acid having the general formula, $R[CH_2CH-(COOR')]_n SCH_2CO_2H(R'$—$C_{1-14}$ alkyl and $n = 5$–$10)(33)$. Another variation involving this complex incorporates titanium lactate as an anchoring agent for resin such as poly(vinyl alcohol)(*34*).

The main chemical variation in the chrome complexes is that the organic portion contains a reactive double bond or an active amino or other reactive nitrogen-containing group. A recent patent(*35*) relating to bonding between synthetic and organic rubbers specifies that the anchoring agent of a Werner complex compound should contain an —SH on the carboxylate group coordinated with the chromium. Once again the double reactivity of the anchoring agent is emphasized as the

Cr portion should be able to react with the —SiOH groups on the glass surfaces. Chromium complexes are usually used with polyester, epoxy, and phenolic resins.

B. Silane Coupling Agents

These compounds have grown in usefulness as coupling agents as organic silicon compounds became commercially available. Once again as with the chrome complexes the silanes must fulfill two basic requirements. They must have the ability to react with the glass surface (—SiOH) to form an —SiOSiR bond, and the R portion of the silane must contain a reactive double bond, amino, reactive nitrogen group, SH group, or epoxy group. In Table II some of the more recent examples of silane uses are listed.

TABLE II

RECENT EXAMPLES OF SILANE APPLICATIONS TO GLASS FIBERS

Compound	Reference
Vinyltris (2-methoxy ethoxy) silane with PV Acetate	Brit. Pat. 848,271, Owens Corning Fiberglass
Vinyl trichlorosilane	*Kunststoffe Plastics*, **13** (1), 2 (1966)
Vinyl silane, i.e., diallyldiethoxysilane	Swed. Pat. 198,775, Allmanna Svenska Elektriska Aktiebolog
Vinyl and NH groups are attached through siloxane groups on other polymers	U.S. Pat. 3,249,461, T. A. TeGrotenhuis
Vinyl triethoxy silane with methyl cellulose or polyacrylate	Jap. Pat. 19,480 (65), Nitto Boseki Co., Ltd.
Vinyl silanol used with film-forming resin	Ger. Pat. 1,019,637, Libbey-Owens-Ford
Trimethyl chlorosilane	*Plasticheskie Massy*, **1966** (2), 27
Poly(methylphenylsiloxanes)	*Plasticheskie Massy*, **1966** (1), 65
Methylpolysiloxane	Brit. Pat. 855,030, General Electric Co.
Monoepoxide–amino alkylsilicon adduct in conjunction with resinous binder	U.S. Pat. 3,259,518, Union Carbide Corp.
Amino silane and a copolymerizable monomer, $RO(CH_2)_4$ or (R-glycidyl)	U.S. Pat. 3,252,825, Owens-Corning Fiberglass
(γ-amino propyl) triethoxy silane in a natural rubber latex containing resorcinol–HCHO resin	Belg. Pat. 669,555, Neth. Appl. 6,512,528, Owens-Corning Fiberglass
Amide of γ-amino propyl triethoxy silane plus dicarboxylic acid	U.S. Pat. 2,919,173, J. P. Stevens

A history of the patent literature as well as a general bibliography on glass–resin interfaces has been compiled by Eakins(*36*). It is claimed that many of the silane coupling agents can be applied in the vapor phase(*37*) but reproducibility is a problem. Also very high temperatures (275°C) seem to be necessary to get the proper interaction. However, this method may grow in importance as technology develops.

The adhesion of resins to silane-treated fibers and fabrics involves consideration of several factors. These include (1) cleanliness of surface, (2) moisture content, (3) reaction temperature, (4) surface tension, and (5) reaction time and pressure. The temperature dependence of the coupling reaction of thermoplastics to silane-treated glass may be limited by the heat distortion or decomposition temperature of the resin as well as by the thermal stability of the silane. The upper limits in molding temperature sometimes exceed the stability of silanes. Plueddemann has studied the silane coupling agents in glass-reinforced polystyrene molded at temperatures between 350–625°F(*38*). A large number of silanes were evaluated with the following general conclusions. Propylsiloxanes show the poorest thermal stability as aliphatics are generally known to have. Their stability is usually adequate for common thermosetting resins but of doubtful usefulness with thermoplastics which require higher molding temperatures. As would be predicted N-phenylaminopropylsiloxane is more stable than aliphatic aminopropylsiloxane. The arylsiloxanes are most stable and vinylsiloxane is almost equivalent. There is danger in such a general discussion of coupling activity versus temperature for often it may be found that effective reactivity is not evident over a range of molding temperatures. It is also possible, particularly in the reinforced plastics systems, that the resins when brought into contact with the glass are of high molecular weight and thus have few reactive sites (*39*) per gram.

An extensive evaluation of coupling agents has been carried out by Englehardt and coworkers(*40*) employing the thermoplastics polypropylene and polystyrene–acrylonitrile. These are two quite different polymers and molding conditions and variables were studied over a range of temperatures. Of the 55 coupling agents evaluated, the best agent for polypropylene–glass fiber was a commercial product described only as a polypropene type. The agents best for the styrene–acrylonitrile polymer were described commercially as an "amino type," γ-hydroxypropyltrimethoxysilane, $HO(CH_2)_3Si(OCH_3)_3$, and bis-(β-hydroxyethyl)aminopropyltriethoxy silane, $(HOCH_2CH_2)_2N$-$(CH_2)_3Si(OC_2H_5)_3$. As might be expected in thermoplastic molding resins, the quantity of glass fiber used for reinforcement and its length and diameter play an important role in determining the final tensile and flexural strengths of the product. About 40% fiber gave a maximum tensile strength for both resins but flexural strength was maximum at 35% fiber for polypropylene and at 45% for styrene–acrylonitrile polymer.

The exact role of the silane coupling agents has been the subject of a large amount of discussion and controversy in recent years. In accounting for their function, the most widely accepted theory suggests that the coupling agent forms covalent bonds to both the glass surface and the resin(*41a*). Wong(*41b*) has discussed the mechanism of coupling by silanes of epoxies to glass fibers. His work involves a size composed of $H_2N(CH_2)_3Si(OC_2H_5)_3$ and a low molecular weight epoxy resin as a film former. The silane forms a monolayer on the glass, a portion of the film former reacts with the silane amino groups and the remainder of the film former reacts with the resin used in preparing the final composite. Sterman and Marsden(*41*) have also studied silane coupling agents for bonding a variety of polymers to glass. Polymers included are both thermosetting and thermoplastic. Islinger, Gutfreund, Maguire, and Olson(*42*), however, were unable to find evidence for the copolymerization of styrene with a vinyl–silane absorbed on a glass. Other theories have been that the interactions at the coupling agent interface are due to hydrogen-bonding phenomenon between silanols in the coupling agent and the glass-surface hydroxyl groups. Still others claim that van der Waals forces at the interface are sufficiently strong to account for the improved strengths in laminates. Rather thorough reviews of the theories on the function of coupling agents have been prepared by Johannson(*43*) and Erickson(*44*).

The determination of distribution of the coupling agents on the glass surface has been a difficult problem because the relative amount of the coupling agent used is exceedingly small. Most researchers have assumed that the agents are distributed uniformly on the glass surface, yet in certain instances electron photomicrographs have shown that coupling agents are sometimes deposited as discrete agglomerates. This indicates high surface tensions between the glass and the coupling agent or subsequent migration of the agent during the heating phases. However, Vogel(*45*) and coworkers suggest that incomplete removal of the glass from the replica for electron microscopy could yield misleading photomicrographs.

A fundamental study of bonding between glass and plastic has been carried out by Chamberlain(*46*) utilizing the following chemistry:

$$\text{—SiOH} + 2AX \rightarrow SiX + A_2 + HX \tag{1}$$

where —SiOH is a silanol group on the glass surface and AX is a chlorine or fluorine carrier;

$$-SiX + RM \rightarrow SiR + MX \qquad (2)$$

where RM is an organometallic and R is an organic radical such as phenyl, or allyl. This sets up a Si–C bond which is not dependent upon an electron-donating atom such as oxygen. Contact angles of the resulting glass surfaces are given for glass with a propyl–butyl surface, a propyl surface, and a fluorinated-butyl surface for a variety of liquids.

V. Glass Fibers in Composites

The polymers which have been used to coat glass fibers or fabrics and those which have been used in the molding industry with glass fibers for fillers or reinforcement fall into two major categories — the thermosetting resins and the thermoplastic resins. (Cellulose derivatives, which are sometimes used in fabric coatings, are considered here as a thermoplastic resin even though the more common ones decompose before or as they melt.) However, a fair treatment of the chemistry involved in utilizing glass fiber necessitates not only references to the resins used in large volumes but some discussion of their many modifications. There are many newer, low-volume, specialty resins but for the most part they are laboratory samples at present. The reinforced-thermoplastics field alone has grown rapidly due to the contribution that glass fibers have made in improving and modifying the physical properties of these polymers. An excellent summary of the thermoplastics, with their physical, thermal, and electrical properties, is available (47) although the effect of glass fibers as a filler or in laminates has not been included.

The glass fiber in reinforced plastics is short staple, roving, filament, or fabric. These give the thermoplastic improved tensile strength, stiffness, dimensional stability, creep resistance, heat resistance, and hardness. The more frequently used resins are polyester, epoxy, nylon, polystyrene, ABS, polycarbonate, polypropylene, acetal, SAN, polysulfone, and HD polyethylene. Such reinforced-plastic materials may have the inherent properties of polymers. They can offer good chemical resistance, are lightweight, have good electrical properties, and can be fabricated in many ways. Acetal resins which are a recent entry in this field show greatly increased stiffness with the addition of glass fibers, reduced coefficient of linear thermal expansion, and excellent creep resistance.

Glass-filled polycarbonates are noted for their improved temperature resistance and very low coefficient of linear thermal expansion as well as low mold shrinkage (0.2–0.4%). This coupled with virtually no water absorption allows moldings to very close tolerances. However, impact strength is reduced.

Polybenzimidazoles (PBI) have been receiving attention as one of the newer polymers which have application in the glass-fiber field as adhesives(48) and laminates(49, 50). These high-temperature-resistant resins naturally present another problem in the glass-fiber field since a very-heat-stable coupling agent is required with polymers useful at 600°F. At the present time cyanophenyl, carboxyphenol, and α-bromotolyl silanes seem to be the best coupling agents for such resins(51).

Epoxy resins which are in common usage in the glass-fiber field depend very much for their properties on the hardener or curing agent that is used with them. Such agents include aromatic amines as dimethylbenzylamine, polymerization products of a diisocyanate and phenylaminomethyl diethoxysilane(52), and imidazoles such as 2-ethyl-4-methylimidazole(53). However, this thermally stable polymer did introduce some new processing problems, for curing temperatures of 700°F may be required as well as thermally stable mold-release agents(54). Cyclopentane dianhydride is a curing agent for high-temperature epoxy resins. The cure can be carried out in steam-heated ovens. The use of a catalyst to improve this system has not been too successful. 2-Ethyl-4-methylimidazole may offer some acceptable curing properties along with bisphenol A and 1,4-butanediol(55). Polyurethanes also require a curing agent such as a diisocyanate; with unsaturated polyesters, allyl isocyanurates and dicumyl peroxide may be used.

The curing of polyester resins has been evaluated by physical test methods. However, neither hardness nor electrical resistivity appeared to be suitable for accurate measurement of the last stages of curing(56). Benzoyl peroxide and methylethyl ketone peroxide are commonly used to catalyze different types of polyester resins. The effect of water on the gel and cure times can vary with the resin, catalyst, promoter, and amount of water added. Water can have a negligible effect at low concentrations, but above 1%, especially in the presence of a cobalt promoter, it can strongly inhibit the gel and cure. On the other hand, water can act as an accelerator but its effective concentration depends also on other variables(57).

The basic standby resins are the phenol–aldehyde and the aminoplast types. Improvements in these have been made by replacing the cations of alkali metals, via ion exchange treatment of the resin, with aluminum and certain other Group-II and -III metals(58), NH_4^+ or amines(59), and urea with ammonium lignosulfonates(60). Progress in thermoplastic-resin usage may be noted in the polyamide work. Schaaf's comparison(*61*) of nylon-6, nylon-6,6, and nylon-12 reinforced with 30% glass fiber show improvement in modulus of elasticity, bending strength, creep resistance, dimensional stability, heat distortion temperature, and shrinkage. These polyamides were chosen to provide a cross section in properties, 6 being ductile, 6,6 hard, and 12 hardly influenced in any way by water or moisture content.

The latest technique for curing resins is by radiation. The kinetics of γ-ray initiated copolymerization of styrene with a low molecular weight unsaturated polyester (derived from maleic anhydride, phthalic anhydride and propylene glycol) have been studied by Burlant and Hinsch (*62*). The formation of gel fractions and the resulting immobility of gel radicals introduce a variation which precludes application of conventional copolymer composition theory. When the disappearance of the monomer and the formation of gel fractions were measured as a function of dose rate, the polymerization rate exhibited linear dependence on intensity and then a region independent of intensity. These workers suggest from their data that the polymerization occurs in discrete volume elements and that chain growth is initiated by hot radicals. The volume of the element in which polymerization is occurring was estimated from experimental data as 3×10 Å.

Further study on curing has employed a 500-keV electron accelerator and mixtures of three vinyl monomers, styrene, ethyl acrylate, and methyl methacrylate, with two unsaturated polyesters(*63*). At five different compositions of styrene–polyester, it has been observed that gel content increases continuously with dose even after the monomer, styrene, apparently has disappeared or reached a level-off concentration. As the monomer content is increased to about 25%, a limit is apparently reached in the amount that will react (or become nonvolatile). In the curing process with low styrene concentration, all of the styrene is reacted with the excess amount of polymer unsaturation. The styrene forms a copolymer with about 2–4 moles of styrene per mole of polymer unsaturation. Thus, at low styrene concentrations copolymer gel and soluble, essentially unreacted, polyester are the two main products. There are three products (copolymer gel, homopolystyrene, and unreacted styrene) at high styrene concentrations due to the limited unsaturation in the polyester. Dose rates in all of these studies were varied from 0.44 Mrad/min up to 100 Mrad/min.

The use of glass fibers as a rubber reinforcement media has been growing significantly. Glass fibers as a tire cord in belted tires show improved impact resistance, cooler running, and longer wear than tires containing a synthetic fiber. These claims are encouraging and as further developments are made in automobile tire construction, the glass fiber will have many desirable properties to contribute.

REFERENCES

1. *Textile Organon*, **40** (12), 202 (1969).
2. W. H. Zachariasen, *J. Amer. Ceram. Soc.*, **54**, 3841 (1932).
3. *Ciba Reviews* (5), 24 (1963).
4. M. B. Volf, "Technical Glasses," Pitman and Sons, London, 1961.
5. J. Hansmann, *Melliand Textilber.*, **47**, 614 (1966).
6. "Textile Fiber Materials for Industry," Owens-Corning Fiberglass Corp., 1964, p. 17.
7. L. Forshaw and P. E. Jellyman, *Research*, **14**, 397 (1961).
8. W. J. Kroenke, *J. Amer. Ceram. Soc.*, **49**, 508 (1966).
9. K. I. Blokh, *Izv. Akad. Nauk SSSR, Neorgan. Materialy*, **2**, 1280 (1966); T. D. Andryukhina, *Steklo i Keram*, **23**, 11, (1966).
10. G. M. Bartenev and R. G. Chernyakov, *Dokl. Akad. Nauk SSSR*, **174**, 800 (1967).
11. G. M. Bartenev and L. K. Izmailova, *Fiz. Tverd. Tela*, **6**, 1192 (1964).
12. N. M. Bobkova and I. A. Trunets, *Steklo i Keram*, **23**, 13 (1966).
13. H. E. Hagy, *Central Glass Ceram. Res. Inst. Bull.*, **13**, 29 (1966).
14. D. L. Hollinger and H. T. Plant, Proc. Ann. Tech. Conf., SPI Reinforced Plast. Div., 21st Chicago, Sect. 13B, 1966.
15. Netherlands Application 6,513,600, N.V. Silenka-Aku-Pittsburgh.
16. U.S. Patent 3,265,479, Owens-Corning Fiberglass Corp.
17. French Patent 1,429,967, Fiberglass, Ltd.
18. Netherlands Application 6,408,881, Owens-Corning Fiberglass Corp.
19. F. J. Lachut, *Am. Dyest. Reptr.*, **48**, 43 (1959).
20. E. L. Lotz, *Am. Dyest. Reptr.*, **48** (4), 50 (1959).
21. British Patent 1,063,162, J. P. Stevens & Co; U.S. Patent 3,382,135; 3,375,155, J. P. Stevens & Co.
22. W. J. Eakins, *Soc. Plastic Eng. Trans.*, **1, 2**, 354 (1961–62).
23. W. J. Eakins, Proc. Ann. Tech. Management Conf., Reinforced Plastics Div., Soc. Plastics Ind., Sec. 10-C, 1962; ibid., Sec. 15-C, 1965.
24. V. B. Tikhomirov, *Dokl. Akad. Nauk SSSR*, **167** (4), 867 (1966).
25. V. Zvonar, *Plaste Kautschuk*, **12** (6), 340 (1965); ibid., **12** (11), 660 (1965).
26. D. I. James, R. H. Norman, and M. H. Stone, *Plastics and Polymers*, **36** (121), 21 (1968).
27. Lieng-Huang Lee, 22nd Ann. Mtg. Reinforced Plastics Div. SPI, 13C, Washington, D. C., 1967.
28. B. M. Vanderbilt and J. J. Jarugelski, *Industr. Eng. Chem.* (*Prod. Res. Develop.*), **1**, 188 (1962).
29. J. T. Englehardt, F. G. Krautz, T. E. Phillips, J. A. Preston, and T. P. Wood, SPI, Reinforced Plastics Div. Conf. Proceedings, 10E, Washington, D. C., 1967.
30. R. C. Horton, *Am. Dyest. Reptr.*, **48** (4), 48 (1959).
31. Edwin L. Lotz, *Am. Dyest. Reptr.*, **48** (4), 50 (1959).
32. Belgian Patent 669,551, Owens-Corning Fiberglass Corp.
33. U.S.S.R. Patent 173,705.
34. U.S. Patent 2,938,812, Owens-Corning Fiberglass Corp.
35. U.S. Patent 3,364,059, Owens-Corning Fiberglass Corp.
36. W. J. Eakins, "Glass/Resin Interface": Patent Survey, Patent List, and General

Bibliography, U.S. Dept. Commerce, Defence Documentation Center, A.D. 609,526, Sept., 1964.

37. U.S. Patent 3,276,853, DeBell and Richardson, Inc.

38. E. P. Plueddemann, Pros., Ann. Tech. Conf., SPI (Soc. Plast. Ind.) 21st Reinforced Plast. Div., Sec. 3D, Chicago, 1966.

39. S. Sterman and J. G. Marsden, Proc. Ann. Tech. Conf. SPI (Soc. Plast. Ind.) 21st Reinforced Plast. Div., Sec. 3A, Chicago, 1966.

40. J. T. Englehardt, F. G. Krautz, T. E. Phillips, J. A. Preston, and R. P. Wood, Proc. Ann. Tech. Conf. SPI (Soc. Plast. Ind.) 22nd Reinforced Plast. Div., Sec. 10E, Washington, 1967.

41a. H. A. Clark, E. P. Plueddemann, Proc. 18th Ann. Tech. and Management Conf., Reinforced Plastics Div., SPI Inc. Sec. 20C, 1963.

41b. Robert Wong, "Conf. on Fundam. Aspects of Fiber Reinf. Plast. Compos., Dayton, Ohio, 1966," eds. R. T. and H. S. Schwartz, Wiley-Interscience, New York, 1968, p. 237.

41c. S. Sterman and J. G. Marsden, "Conf. on Fundam. Aspects of Fiber Reinf. Plast. Compos., Dayton, Ohio, 1966," eds. R. T. and H. S. Schwartz, Wiley-Interscience, New York, 1968, p. 245.

42. J. S. Islinger, K. Gutfreund, R. G. Maguire, and O. G. Olson, WADC Tech. Report 59–600, Part I, March 1960.

43. O. K. Johannson, *et al.*, AFML-TR-65-303, Pt. 1, September 1965.

44. Porter W. Erickson, NOLTR-63-253, November 1963.

45. G. E. Vogel, O. K. Johannson, F. O. Stack, and R. M. Fleischmann, Proc. Ann. Tech. Conf., SPI (Soc. Plast. Ind.) 22nd Reinforced Plast. Div., Sec. 13B, Washington 1967.

46. D. L. Chamberlain, Jr., SPI (Soc. Plast. Ind.) 22nd Reinforced Plastics Div., Sec. 13E, Washington, 1967.

47. J. Brandrup and E. H. Immergut, "Polymer Handbook," Wiley-Interscience, New York, 1966, Chap. 9.

48. J. R. Hill, *Adhesives Age*, **9** (8), 32 (1966).

49. British Patent 1,034,293, Whittaker Corp.

50. H. A. Mackay, *Mod. Plastics*, **43** (5), 149 (1966).

51. E. P. Plueddemann, SPI (Soc. Plast. Ind.) 22nd Reinforced Plastics Div., Sec. 9A, Washington, 1967.

52. French Patent 1,401,823.

53. A. Farkas and P. F. Strohm, *J. Appl. Poly. Sci.*, **12**, 159 (1968).

54. Roland Ried and T. J. Reinhart, Jr., 21st Annual Meeting Reinforced Plastics Div., SPI, Sec. 7B, 1966.

55. Ross Van Volkenburgh and W. C. Johnson, 21st Annual Meeting Reinforced Plastics Div., SPI, Sec. 1A, 1966.

56. G. S. Learmonth, F. M. Tomlinson, and J. Czerski, *J. Appl. Poly. Sci.*, **12**, 403 (1968).

57. J. D. Malkemus, 21st Annual Meeting Reinforced Plastics Div., SPI, Sec. 7D, 1966.

58. British Patent 842,634, Owens-Corning Fiberglass Corp.

59. British Patent 815,414, Owens-Corning Fiberglass Corp.

60. U.S. Patent 3,336,185, Pittsburgh Plate Glass Co.

61. S. Schaaf, SPI (Soc. Plast. Ind.) 22nd Reinforced Plastics Div., SPI, Sec. 10B, Washington, (1967).

62. W. Burlant and J. Hinsch, *J. Poly. Sci., Part A*, **2**, 2135 (1964); *ibid.*, **3**, 3587 (1965).

63. A. S. Hoffman and D. E. Smith, *Mod. Plastics*, **43** (10), 111 (1966).

AUTHOR INDEX

Numbers in parentheses are reference numbers and indicate that an author's work is referred to although his name is not cited in the text. Numbers in italics show the page on which the complete reference is listed.

Haley, A. R., 77(25), 78, *88*
Haly, A. R., 74(14), *88*
Hamburger, C. J., 61, *65*
Hamilton, C. E., 149(19), *171*
Hammond, G. S., 149(19), *171*
Hanley, S., 3(3), *29*
Hansen, R. H., 150(11), *171*
Hansmann, J., 185(5), *198*
Harbrink, P., 58(3), *64*
Harper, R. J., Jr., 12(18), *29*
Harrap, B. S., 72(11), *88*
Hashiya, S., 158(21), *172*
Hata, Dunio, 36(2), *52*
Hatton, J. R., 158(21), *172*
Hausel, H., 40(14), *52*
Hawkins, W. L., 150(12), 152, *171*
Haworth, S., 14(31), *29*
Hay, J. N., 117(25), *126*
Hayahara, T., 116(21), *126*
Hayakawa, K., 61(23), *65*
Hayashi, S., 106(80, 81), *112*
Healy, E. M., 97(29), *111*
Heap, S. A., 99(46), *111*
Hebeish, A., 26(77), *31*
Hepp, J., 11, *29*
Heterington, P. W., 7(11), *29*
Heumann, W., 41(19), *52*
Hewlett, C., 50(43), *53*
Hill, J. R., 196(48), *199*
Hill, R., 137, *142*
Hinojosa, O., 26(71), *30*
Hinsch, J., 197, *199*
Hiroi, I., 120(35), *126*
Hoffman, A. S., 162, *172*, 196(63), *199*
Hoffmann, G., 40(14), *52*
Hollinger, D. L., 187, *198*
Hoiness, D. E., 13(27), *29*
Holker, J. R., 84(63), *89*
Holmes, F. H., 99(46), 102(54), *111*
Hopff, H., 94(8), *110*
Horio, M., 71(2), 78, 87, *88*
Horton, R. C., 190(30), *198*
Howard, G. J., 131(5), *142*
Howard, W. H., 94, *110*
Huang, R. M., 26(79), *31*
Hybart, F. J., 93(3), *110*
Hyo, M., 64(42), *65*

I

Ikeda, M., 117(22), *126*

Imoto, M., 106(81), *112*
Ingram, P., 87(89, 91), *90*
Imai, Y., 26(70), 27(80), *30*, *31*
Immergut, E. H., 195(47), *199*
Imoto, M., 63(32), *65*
Inoue, Y., 61(22), 64(42), *65*
Ishikawa, I., 102(55), *111*
Ishizuka, A., 162(29), *172*
Ishizuka, O., 117(26), *126*
Islinger, J. S., 194, *199*
Ivanov, N. N., 121(52), *126*
Iwakura, Y., 26(70), 27(80), *30*, *31*
Izmailova, L. K., 187(11), *198*
Izuka, Y., 28(87), *31*

J

Jabloner, H., 149(18), *171*
Jackson, D. L. C., 85(67), *89*
Jackson, J. B., 158(21), *172*
Jackson, R. A., 161(24), *172*
Jackson, W. J., Jr., 177(9), *182*
James, D. I., 189(26), *198*
Jansch, M., 103(68), *112*
Janssen, H. J., 13(27), *29*
Jarugelski, J. J., 190(28), *198*
Jeffries, R., 13, *29*
Jin, C. R., 9, *29*
Joarder, G. K., 13(25), *29*
Joh, Y., 114(9), *125*
Johannessen, B., 22(50), *30*
Johannson, O. K., 194, *199*
Johansen, G., 94(5), *110*
Johnson, D. J., 72, *88*
Johnson, W., 117(27), *126*, 196(55), *199*

K

Kachan, A. A., 104(69), 106(75–78), *112*
Kaeppner, W. M., 41, *52*
Kagaku, K., 162(29), *172*
Kaizerman, S., 26(76), *30*
Karar, S. K., 136(18), *142*
Kakinuma, Toshiko, 28(87), 51(45), *31*, *53*
Kalinichenko, A. M., 106(77), *112*
Kamalou, S., 121(52), *126*
Konnova, N. F., 122(60), *127*
Kantouch, A., 81(42, 44), *88*
Karlin, E. R., 97(26), *111*
Karpov, U. L., 106(84), *112*
Kase, M., 23(53), *30*

SUBJECT INDEX

A

Acetal, 10, 175, 195
Acrolein, 121
Acrylates
  acrylic, 114, 121, 124, 125
  cellulose acetate, 61, 64
  cotton, 23, 26, 28
  glass, 189, 191, 197
  polyamide, 97, 106
  polyethylene terephthalate, 141
  polyolefin, 158, 160, 162, 163, 165, 166
  rayon, 50, 51
  wool, 85, 86
Acrylics
  antistats, 123
  color, 117
  comonomers, 114, 116, 122–124
  crosslinking, 117, 120, 121
  definition, 113
  degradation, thermal, 117, 119–121
  dyeability, 123, 124
  effect of
    heat, 117
    oxygen, 117
  flameproofing, 122
  grafting, 122

production, 113
properties
  chemical, 116
  physical, 114, 116, 117
spinning, 115, 116
  solvents, 115
stablizers, 120
structure
  chemical, 113
  physical, 115, 116
tacticity, 114
Acrylonitrite
  acrylic, 114, 125
  cellulose acetate, 61
  cotton, 26, 27
  polyamide, 105, 106
  polyolefin, 158–161
  rayon, 50
Allophanate, 175
Aziridine, 10, 87, 138, 165

B

Benzoquinone, 84, 151, 156
Benzoylation, 103, 104
Biuret, 175